Advanced Sciences and Technologies for Security Applications

Indexed by SCOPUS

The series Advanced Sciences and Technologies for Security Applications comprises interdisciplinary research covering the theory, foundations and domain-specific topics pertaining to security. Publications within the series are peer-reviewed monographs and edited works in the areas of:

- biological and chemical threat recognition and detection (e.g., biosensors, aerosols, forensics) - crisis and disaster management - terrorism - cyber security and secure information systems (e.g., encryption, optical and photonic systems) - traditional and non-traditional security - energy, food and resource security - economic security and securitization (including associated infrastructures) - transnational crime - human security and health security - social, political and psychological aspects of security - recognition and identification (e.g., optical imaging, biometrics, authentication and verification) - smart surveillance systems - applications of theoretical frameworks and methodologies (e.g., grounded theory, complexity, network sciences, modelling and simulation)

Together, the high-quality contributions to this series provide a cross-disciplinary overview of forefront research endeavours aiming to make the world a safer place.

The editors encourage prospective authors to correspond with them in advance of submitting a manuscript. Submission of manuscripts should be made to the Editor-in-Chief or one of the Editors.

More information about this series at https://link.springer.com/bookseries/5540

Stanislav Abaimov · Maurizio Martellini

Machine Learning for Cyber Agents

Attack and Defence

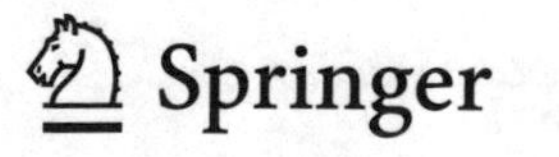

Springer

Stanislav Abaimov
University of Bristol
Bristol, UK

Maurizio Martellini
Fondazione Alessandro Volta (FAV)
University of Insubria
Como, Italy

ISSN 1613-5113 ISSN 2363-9466 (electronic)
Advanced Sciences and Technologies for Security Applications
ISBN 978-3-030-91587-2 ISBN 978-3-030-91585-8 (eBook)
https://doi.org/10.1007/978-3-030-91585-8

This Springer imprint is published by the registered company Springer Nature Switzerland AG
The registered company address is: Gewerbestrasse 11, 6330 Cham, Switzerland

Contents

Abbreviations

AI	Artificial intelligence
ANN	Artificial neural networks
ASIC	Application-specific integrated circuit
AWS	Autonomous weapon systems
BMS	Building management system
CPU	Central processing unit
DARPA	Defence advanced research project agency
DBNN	Deep belief neural network
DNN	Deep neural network
DPI	Deep packet inspection
FPGA	Field-programmable gate array
GPU	Graphics processing unit
HIDS	Host-based intrusion detection system
HMI	Human–machine interface
HMM	Hidden Markov models
ICRS	International Committee of the Red Cross
ICS	Industrial control system
IDS	Intrusion detection system
ITU	International Telecommunication Unit
LAWS	Lethal autonomous weapon systems
NIDS	Network intrusion detection system
OCR	Optical character recognition
PLC	Programmable logic controller
RFID	Radio-frequency identification
SCADA	Supervisory control and data acquisition
TPU	Tensor processing unit
UN	United Nations
UNESCO	United Nations Educational, Scientific and Cultural Organization
UNIDIR	United Nations Institute for Disarmament Research
WHO	World Health Organization

Of note: The term "Machine Learning" will be used in capital letters with reference to the field of science, and with uncapitalized letters when referring to the technical process.

Disclosure Statement

The views expressed in this book are those of the authors only and may not necessarily be shared by some of the experts in Artificial Intelligence and Machine Learning.

The evaluations and reviews are based on the experimental results and theoretical explorations of the authors. The authors exclude any responsibility for losses, damages, costs, expenses, arising directly or indirectly from using this publication and any information or material contained in it.

List of Figures

List of Tables

Chapter 1
Introduction

Artificial Intelligence (AI) entered the information and cyber domains, military affairs and research and development of cognitive systems, driven by the necessity to have human tasks implemented by a digital computer, or computer-controlled robot. Machine Learning, used interchangeably with the term "AI", has become a new powerful tool able to collect massive volumes of data from all areas of human life, rapidly process it and make rational decisions thus allowing to increase the performance efficiency, safety and security, and drive technological progress further.

The 2015 Report of the Group of Governmental Experts on Development in the Field of Information and Telecommunications highlights: "Few technologies have been as powerful as information and communications technologies in reshaping economies, societies and international relations. Cyberspace touches every aspect of our lives. The benefits are enormous, but these do not come without risk."[1] Notable scientists and futurologists warn the world about the AI threats to humanity, technical experts and decision makers sign open letters appealing to ban the AI use for military purposes.[2]

[1] United Nations. Group of Governmental Experts on Developments in the Field of Information and Telecommunications in the Context of International Security, Note by the Secretary General, 2015, http://www.un.org/ga/search/view_doc.asp?symbol=A/70/174.

[2] Autonomous Weapons: An Open Letter from AI and Robotics Researchers, 28 July 2015, https://futureoflife.org/open-letter-autonomous-weapons/.

S. Abaimov and M. Martellini, *Machine Learning for Cyber Agents*,
Advanced Sciences and Technologies for Security Applications,
https://doi.org/10.1007/978-3-030-91585-8_1

Opinion leaders about Artificial Intelligence

We cannot predict exactly how the technology will develop or on what timeline. Nevertheless, we must plan for the possibility that machines will far exceed the human capacity for decision making in the real world. What then?[3]

Stuart J. Russell

The development of full artificial intelligence could spell the end of the human race...It would take-off on its own and redesign itself at an ever-increasing rate... Humans, who are limited by slow biological evolution, couldn't compete and would be superseded.[4]

Stephen Hawking

Progress in robotics and artificial intelligence (AI) should be oriented *towards respecting the dignity of the person and of Creation.*[5]

Pope Francis

I think we should be very careful about artificial intelligence. If I were to guess like what our biggest existential threat is, it's probably that. So we need to be very careful with the artificial intelligence. Increasingly scientists think there should be some regulatory oversight maybe at the national and international level, just to make sure that we don't do something very foolish.[6]

Elon Musk

I am in the camp that is concerned about super intelligence. First the machines will do a lot of jobs for us and not be super intelligent. That should be positive if we manage it well. A few decades after that though the intelligence is strong enough to be a concern. I agree with Elon Musk and some others on this and don't understand why some people are not concerned.[7]

Bill Gates

---

[3] Russell [1].

[4] Rory Cellan-Jones, Stephen Hawking warns artificial intelligence could end mankind, BBC, 2 December 2014, https://www.bbc.com/news/technology-30290540.

[5] Pope's November prayer intention: that progress in robotics and AI "be human", Vatican News, November 2020, https://www.vaticannews.va/en/pope/news/2020-11/pope-francis-november-prayer-intention-robotics-ai-human.html.

[6] Science Focus, The Home of the BBC Science Focus Magazine, https://www.sciencefocus.com/future-technology/should-we-be-worried-about-ai/.

[7] Science Focus, The Home of the BBC Science Focus Magazine, 17 February 2017, https://www.sciencefocus.com/future-technology/should-we-be-worried-about-ai/.

AI promises to bring both enormous benefits, in everything from health care to transportation, and huge risks.[8]

Paul Scharre

AI could destroy humanity as we know it.[9]

Nick Bostrom

Machines and robots that outperform humans across the board could self-improve beyond our control—and their interests might not align with ours.[10]

(*Nature, 2016*)

The expectations from the AI potential capacities have created multiple concerns, debates, and conspiracy theories. In the popular culture, the emerging AI has been always seen as a threat to humanity in one way or another. To illustrate the point, it is enough to open a dedicated page in Wikipedia, that lists existential risks from AI.[11]

The concerns about the inevitable threats typically originate from the yet to be evolved *General AI*, arguably capable to learn autonomously from various sources, adapting to complex environment and making its own decisions. What is the reality nowadays is a narrower aspect of AI, a machine learning algorithm that learns from a set of specified data to implement specific action. **This book focuses on the technical aspects of this so-called** *Narrow* **or** *Specialised AI, Machine Learning in particular,* **its concepts, operation, application, limitations and future challenges**. Thus, from here on, the term AI will refer to the *Machine Learning* techniques, unless specified otherwise.

The most prevalent approach to Narrow AI is machine learning, which involves statistical algorithms that replicate human cognitive tasks by deriving their own procedures through analysis of large training data sets. During the training process, the computer system creates its own statistical model to accomplish the specified task in situations it has not previously encountered.[12]

Artificial Intelligence and National Security, 2020

[8] Paul Scharre, Killer Apps. The Real Dangers of an AI Arms Race. May/June 2019, Foreign Affairs, https://www.foreignaffairs.com/articles/2019-04-16/killer-apps.

[9] Artificial Intelligence, Full debate, Doha Debates, 3 April 2019, 43:07, https://www.youtube.com/watch?v=ffG2EdYxoCQ.

[10] Anticipating artificial intelligence, Nature 532, 413, 28 April 2016, https://doi.org/10.1038/532413a, https://www.nature.com/articles/532413a.

[11] Existential risk from artificial general intelligence, Wikipedia, accessed 2 May 2021, https://en.wikipedia.org/wiki/Existential_risk_from_artificial_general_intelligence.

[12] Daniel S. Hoadley, updated by Kelley M. Sayler, Artificial Intelligence and National Security, 10 November 2020, Congressional Research Service Report, United States, p. 2, https://fas.org/sgp/crs/natsec/R45178.pdf.

Machine Learning is a field of Artificial Intelligence, and a cross-discipline between computer science and data science. The **learning process** consists of data analysis, generation of patterns or models, their structuring and representation for taking further decisions and actions—either by humans or machines themselves. Machines learn from the data they receive either through data specialists or collecting it through other sources, e.g. data streams in the internet, sensors, etc. As opposed to conventional computational algorithms, machine learning derives formulas from data, and can adjust or even change models, and subsequently decisions.

Machine learning is commonly applied for the approximation of the complex representations of the natural world in scientific research, including space exploration. It already plays an important role in the analysis of large volumes of data from the observations of planets and spacecraft telemetry during space missions. For example, the data from Mars rovers is already transmitted using machine learning methods, which also enable there self-driving capabilities.

Applications trained with machine learning are already widely used in all areas of our life, even though people may not be aware of it. These intelligent devices process languages and recognize images, make analysis and facilitate strategic decision-making, adjust prices depending on demand, detect frauds, play games, etc. Machine learning applications are being widely added to smartphones, tablets, and personal computers as digital assistants and on-line chatbots—Watson (IBM), Alexa (Amazon), Siri (Apple). They are used in fitness trackers, in Google Home and Amazon's Echo, cameras and smart home appliances, security and parking systems, etc. Machine learning algorithms predicting our wishes from the collected data are widely used by companies like Amazon, Netflix, LinkedIn, Facebook suggesting goods that people may wish to buy, books to read, films to watch, places to visit. Machine learning enables creation of music, writings and paintings, it synthesises voices and generates art styles. Online and smartphones applications can produce videos replacing faces of real characters with those taken from any photos.[13] Those apps are used for entertainment purposes, but they can also be used for malicious goals. Luckily, the researchers are already addressing this threat and developing software that is able to detect video manipulation.

The ongoing race for consumer attention has turned the machine learning industry into aggressive contest between technological giants, such as Ali Baba, Amazon, Baidu, Google, IBM, Microsoft, Tencent, and others. They compete in promoting innovations, collecting and processing data, generating specific data sets, in building, training and deploying customized models, offering augmented services through the whole AI systems lifecycle.

[13] Cauduro, A. (2018) Live Deep Fakes—you can now change your face to someone else's in real time video applications. Medium. Available from: https://medium.com/huia/live-deep-fakes-you-can-now-change-your-face-to-someone-elses-in-real-time-video-applications-a4727e06612f.

Estimates suggest that by 2030, there will be 125 billion devices connected to the internet, and 90% of individuals older than 6 will be online.[14]

Sweeping digitalisation and the growing use of AI in our daily life requires its **adaptation to non-experts**, people who need additional guidance for the safe use of the AI-enhanced devices. This adaptation process should include raising awareness on the new technology application, sharing technical and security knowledge, public participation in real-life testing scenarios, development of teaching modules. It would be beneficial to all of us if we know more about this new technology and take well informed decisions when using it.

Internet of Things (IoT) and Industry 4.0

IoT is a general term referring to smart gadgets and any devices connected via Internet and enhanced with digital intelligence/sensors. They can be controlled remotely either by humans or other devices as they are able to communicate with each other without human interference, in the machine-to-machine mode, taking the data from and sending it to the Internet. They facilitate our life being smart enough to collect and analyse data in real time and "intuitively" responding to our needs to provide services required. They provide feedback on the work of the controlled devices, inform us about the needs for their maintenance, etc. They are applied in children's toys, in home appliances and wellness products, in vehicles and their maintenance, cities and control of environment; they may be of a general use (e.g., for security) or industry specific (health devices, Industrial Control Systems); they can be switched on and off with a smartphone or through installed commands.

At the same time, they are threatening our personal privacy, can spread fake news and manipulate public opinion; they contain software flaws and are vulnerable to cyber attacks. Their most vulnerable parts are *routers, webcams, applications.*

The success of AI applications led to *the fourth industrial revolution* related to the IoT implementation—**Industrial Internet of Things (IIoT)**—in industries and businesses. They are both benefiting from the advantages of intelligent technologies and modern hardware—wireless communication, super sensitive data collection devices, analysis of big data, cloud solutions, etc. However, complex emerging technologies contain multiple cyber vulnerabilities that can affect supply chains at any level, and result in extensive damages.

[14] Tania Lațici, Cyber: How big is the threat?, European Parliamentary Research Service (EPRS), Members' Research Service PE 637.980, July 2019, at https://www.europarl.europa.eu/RegData/etudes/ATAG/2019/637980/EPRS_ATA(2019)637980_EN.pdf.

The industry 4.0 market is projected to grow from USD 64.9 billion in 2021 to USD 165.5 billion by 2026.[15]

Machine learning has boosted following technological advancements that decreased costs of computer chips (for example, Radio-frequency identification (RFID) tags that need less power, support wireless communication), development of powerful processors, advanced sensors, wireless (cellular) networks, broadband internet and cloud computing. Direct communication between electronic devices has decreased costs of services, and their connection to the Internet generates massive datasets for training high-performing AI models; powerful processors analyse this data and run algorithms. Cloud computing allowed to access the necessary storage and processing capacities. The new generation of **algorithms** are more effective in use, able to simultaneously perform extensive operations that speeds up learning. Many **datasets** and tools have become available in open-sources and include programming languages and libraries, analysis tools, software development frameworks, pretrained models, etc. Developed by experts, they can be used as a whole or modified to the needs.

The on-going industrial revolution differs from the previous one by its limitless evolving capacities—both in time and dimension. The enabling factors are not only human beings, but also the smart technologies themselves able to produce their more and more intelligent versions.

IoT prognosis

Tech analyst company IDC predicts that in total there will be 41.6 billion connected IoT devices by 2025, or "things." It also suggests industrial and automotive equipment represent the largest opportunity of connected "things," but it also sees strong adoption of smart home and wearable devices in the near term.

Another tech analyst, Gartner, predicts that the enterprise and automotive sectors will account for 5.8 billion devices this year, up almost a quarter on 2019. Utilities will be the highest user of IoT, thanks to the continuing rollout of smart meters. Security devices, in the form of **intruder detection and web cameras will be the second biggest use of IoT devices**. Building automation—like connected lighting—will be the fastest growing sector, followed by automotive (connected cars) and healthcare (monitoring of chronic conditions). **[...].**

[15] Industry 4.0 Market by Technology and Geography—Global Forecast to 2026, June 2021, https://www.reportlinker.com/p04920966/Industry-4-0-Market-by-Technology-Vertical-Region-Global-Forecast-to.html?utm_source=GNW.

Worldwide spending on the IoT was forecast to [...] pass the $1 trillion mark in 2022.[16]

Steve Ranger, 2018

The Internet of Things (IoT) market size was USD 250.72 billion in 2019 and is projected to reach USD 1,463.19 billion by 2027.[17]

Fortune Business Insights, April, 2021

With Machine Learning being an emerging and rapidly developing field, its application in cyber security is also increasing. Estimates indicate that the market for AI in cybersecurity will grow from 1 billion USD in 2016 and 3.92 billion USD in 2017[18] to a 34.8 billion USD net worth by 2025,[19] and further to 101.8 billion USD in 2030.[20] This has already brought feasible advantages in the social, economic and political life, but also new threats. There are some strong considerations to keep in mind while working with machine-learning methodologies or analysing their impact.

IoT vulnerabilities

Over 100,000 internet-connected security cameras contain a "massive" security vulnerability that allows them to be accessed via the open web and used for surveillance, roped into a malicious botnet, or even exploited to hijack other devices on the same network.

Representing yet more Internet of Things devices that are exposed to cyber-attackers, vulnerabilities have been uncovered in two cameras in Chinese manufacturer Shenzhen Neo Electronics' NeoCoolCam range.

Researchers at Bitdefender say the loopholes mean it's trivial for outsiders to remotely attack the devices and that 175,000 of the devices are connected

[16] Steve Ranger, "What is the IoT? Everything you need to know about the Internet of Things right now," ZDNet.com, August 21, 2018, https://www.zdnet.com/article/what-is-the-internet-of-things-everything-you-need-to-know-about-the-iot-right-now/.

[17] The global Internet of Things market is projected to grow from $381.30 billion in 2021 to $1854.76 billion in 2028 at a CAGR of 25.4% in the 2021–2028 period..., May 2021, https://www.fortunebusinessinsights.com/industry-reports/internet-of-things-iot-market-100307.

[18] AI in Cybersecurity Market by Offering (Hardware, Software, Service), Technology (Machine Learning, Context Awareness, NLP), Deployment Type, Security Type, Security Solution, End-user, and Geography—Global Forecast to 2025, https://www.marketsandmarkets.com/market-reports/ai-in-cybersecurity-market-224437074.html, accessed on 20 March 2021.

[19] Taddeo, M., McCutcheon, T., & Floridi, L. (2019). Trusting artificial intelligence in cybersecurity is a double-edged sword. Nature Machine Intelligence, 1(12), 557–560. https://doi.org/10.1038/s42256-019-0109-1.

[20] AI in Cyber Security Market Research Report: By Offering, Deployment Type, Technology, Security Type, Application, End User—Global Industry Analysis and Growth Forecast to 2030, August 2020.

to the internet and vulnerable. Between 100,000 and 140,000 are detectable via the Shodan IoT device search engine alone.

The easy online availability and low cost—some models are available for under £30 ($39)—of Shenzhen products means the NeoCoolCam devices have spread around the world. [...].

The two cameras studied, the iDoorbell model and NIP-22 model, contain several buffer overflow vulnerabilities, some even before the authentication process. The flaws can be used for remote execution on the device—the attacker doesn't even need to be logged in, even just the attempt at a login can provide access. [...].

The camera web server exploit stems from a vulnerability in the HTTP service triggered by the way the application processes the username and password information at login.[21]

The **cyber world** has been both enhanced and endangered by AI. On one side, the performance of the existing security services has been improved, and new tools created. On the other hand, it generates new cyber threats both through enhancing the attacking tools and through its own imperfection and vulnerabilities. The new AI-based attacks and malware like DeepHack (framework), DeepLocker (ransomware) or Deep Exploit (exploitation kit) can be named as examples. Open-source tools accessible to anyone, e.g. Tensorflow, Keras, SciKit, DeepSpeech, increase these threats allowing even people with the minimal technical knowledge to assemble cyber arms and hacking devices.

Botnet detection

Botnet is a combination of multiple bots—malicious programs acting as a command receiver on an infected system—and a Command and Control program (C&C), controlled by a malicious actor.

Botnets are detected using honeypot analysis, communication signature analysis, and behavioural analysis. Behavioural analysis is already carried out using machine learning, specifically deep learning.[22]

Popular C&Cs use specific network ports and can be detected via active mass scanning of a single port, using tools like *nmap* and *masscan*.

The **AI-enhanced weapons**, though raising the defence capacities to a higher level, are bringing existential danger. Among them are self-targeting missiles and

[21] Danny Palmer, 175,000 IoT cameras can be remotely hacked thanks to flaw, says security researcher, 31 July 2017, ZDNet, https://www.zdnet.com/article/175000-iot-cameras-can-be-remotely-hacked-thanks-to-flaw-says-security-researcher/.

[22] Survey on Botnet Detection Techniques: Classification, Methods, and Evaluation, 2020, https://www.hindawi.com/journals/mpe/2021/6640499/.

self-learning unmanned ground vehicles with the built-in automated and autonomous functions, e.g., "Friend-or-Foe" identification. The responsibility for the life of civilians, proportionality, protection of the wounded and sick, universal dignity and other humanitarian principles codified in the Geneva and the Hague conventions are challenged when the life and death decisions are conferred to the artificial intelligence of military robots.

The current state of the AI research and its functionality show that it is too early for the complex and yet imperfect machine learning algorithms to be trusted and granted full or even partial unsupervised autonomy. The **training** stage is a complex and computationally intense process, that typically requires skilled experts with specific knowledge in computer and data sciences, mathematics, etc. AI-systems need a large amount of carefully selected and processed data, task-specific algorithms, optimisation for hardware, evaluation in contained and field environments, maintenance, routine upgrade and adjustments. The **hardware** able to meet the requirements of the evolving AI programmes and to ensure a higher level of security, needs to constantly grow in power, providing adequate storage capacities, ensuring comprehensive data collection from sensors, distributing unhindered signals to other devices, while being regularly checked for functionality and potential vulnerabilities. Human supervision is essential through the whole life cycle of the AI-enhanced devices.

Machine learning and critical infrastructures

In critical infrastructures, machine learning is used for security of industrial control systems (ICS) responsible for safety, stability and real-time high-quality service, e.g. addressing voltage fluctuations. Decisions of control systems should be precise, reliable and predictable, and ICS should be protected from cyber attacks 24/7.

Though offering round the clock services and high speed of reaction, at this stage of their development the machine learning-based security models are not yet mature enough to be trusted with full autonomy in high-risk environment and intrusion detection. Being adaptive to networks and able to detect anomalies, they require high volume of training data, sophisticated software and hardware, they produce multiple false positive signals and are vulnerable to attacks.[23] Human supervision is needed to ensure the required high level of ICS safety and security. However, the issue of the necessary but not over-exercised level of this "meaningful" human control and percentage of affordable errors as compared to human operators remains open.

[23] JASON Study, Perspectives on Research in Artificial Intelligence and Artificial General Intelligence Relevant to DoD, The MITRE Corporation, JSR-16-Task-003, January 2017, https://fas.org/irp/agency/dod/jason/ai-dod.pdf.

The use of the AI-systems at the governmental level, in e-diplomacy and international relations, raises the issue of the AI integration and adaptation to the highest level of **global governance**, and requires global concerted action based on mutually agreed technical standards, legal norms, ethical principles. The question about the amount and level of tasks allowable for implementation by machines remains in discussion and inspire debates at local, national and international levels. The appeals for the AI leadership are heard from the heads of governments stressing its importance for economic development and national security. Intergovernmental organizations convene multilateral conferences and summits, create high-level working groups of experts to discuss a variety of AI-related subjects, from responsibility and ethics to big data and autonomy, the impact of high technologies on the future of the planet.

Global competition for AI leadership

Russia

Artificial intelligence is the future, not only for Russia, but for all humankind. … It comes with colossal opportunities, but also threats that are difficult to predict. Whoever becomes the leader in this sphere will become the ruler of the world.[24] *(Vladimir Putin, President of the Russian Federation).*

China

In 2017, China adopted the "New Generation Artificial Intelligence Development Plan" with an ambitious goal to become a leading power in AI by 2030, "achieving visible results in intelligent economy and intelligent society applications, and laying an important foundation for becoming a leading innovation-style nation and an economic power"[25]

USA

Ceding leadership in developing artificial intelligence to China, Russia, and other foreign governments will not only place the United States at a technological disadvantage, but it could have grave implications for national security.[26]

[24] 'Whoever leads in AI will rule the world': Putin to Russian children on Knowledge Day, Russia Today, 1 September 2017, https://www.rt.com/news/401731-ai-rule-world-putin/.

[25] China State Council, "A Next Generation Artificial Intelligence Development Plan," July 20, 2017, translated by New America, https://www.newamerica.org/documents/1959/translation-fulltext-8.1.17.pdf.

[26] U.S. Congress, Senate Subcommittee on Space, Science, and Competitiveness, Committee on Commerce, Science, and Transportation, Hearing on the Dawn of Artificial Intelligence, Statement of Senator Cruz, 114th Congress, Second session, November 30, 2016, U.S. Government Publishing Office, Washington, 2017, p. 2, https://www.govinfo.gov/content/pkg/CHRG-114shrg24175/pdf/CHRG-114shrg24175.pdf.

France

In March 2018, the President of France announced measures "allowing France to take a place of one of the world leaders in AI"[27] presenting the national strategy at the "AI for Humanity" conference. "Turning the country into an AI leader would allow France to use AI for the public good and ensure that a "Promethean" promise doesn't become a "dystopia," he said."[28]

ASEAN

The ASEAN Digital Masterplan 2025 envisions "ASEAN as a leading digital community and economic bloc, powered by secure and transformative digital services, technologies and ecosystem".[29]

European Union

It is in the Union interest to preserve the EU's technological leadership and to ensure that Europeans can benefit from new technologies developed and functioning according to Union values, fundamental rights and principles.[30]

The state-of-the-art technologies are being developed and programmed to help humanity make informed rational decisions, that could ensure equal benefits, safety and security for each and every one of us.

1.1 Motivation

Currently, it is virtually impossible to stress enough the impact of machine learning on the modern world in all the areas, cyber security being only one of them. It has captured a world-wide academic community attention and is considered by multiple disciplines, i.e. mathematics, physics, political science, philosophy, psychology, biology, etc. On one hand, this multidisciplinary approach enriches scientific debates,

[27] Macron veut faire de la France un leader de l'intelligence artificielle, Reuters, 26 March 2018, https://www.reuters.com/article/france-intelligence-artificielle-idFRKBN1H22CX-OFRTP; see also Rapport Villani, 2018, https://www.aiforhumanity.fr/pdfs/9782111457089_Rapport_Villani_accessible.pdf.

[28] Tania Rabesandratana, Emmanuel Macron wants France to become a leader in AI and avoid 'dystopia', 30 March 2018, https://www.sciencemag.org/news/2018/03/emmanuel-macron-wants-france-become-leader-ai-and-avoid-dystopia.

[29] The ASEAN Digital Masterplan 2025, p.4, https://asean.org/storage/ASEAN-Digital-Masterplan-2025.pdf.

[30] Regulation of The European Parliament and of the Council Laying Down Harmonised Rules On Artificial Intelligence (Artificial Intelligence Act) and Amending Certain Union Legislative Acts, Brussels, 21.4.2021, at https://eur-lex.europa.eu/resource.html?uri=cellar:e0649735-a372-11eb-9585-01aa75ed71a1.0001.02/DOC_1&format=PDF.

but on the other hand, deviating from the technical expertise in understanding the underlying mechanisms and real state of the art, it may be overly misleading. Technical experts, researchers and developers, should be more proactive in raising awareness and addressing both positive and negative sides of the machine learning application, so that informed decisions would be taken timely.

Accurate forecasting and statements regarding AI cannot be realistic without an in-depth technical background and knowledge on the machine learning.

Why is this book needed? With the abundance of the confusing information and lack of trust in the wide applications of the AI-based technology, **it is detrimental to have a book, that would explain, through the eyes and perception of the cyber security expertise, why and at what stage the emerging powerful technology of machine learning can and should be mistrusted, and how to benefit from it while avoiding life threatening consequences**. In addition, it brings to the light another very sensitive area—**application of machine learning for offence, that is widely misunderstood, poorly represented in the academic literature and requires a special expert attention.**

1.2 Aim

This book **aims** at contributing to the conceptualisation and operationalisation of the term "machine learning", through reviewing its constituting elements, functionality and current status. It will present logically structured and referenced **modern practices of machine learning applications in cyber security**, their future challenges, and address major global debates around the use of the AI technology, including for offensive purposes, and its integration into our life while ensuring equal benefits to each and every one of us.

This book, bridging the gap between technical experts and non-technical decision makers, can be used as a supplementary course reading, a reference book, or a reading material for technical experts and general public, seeking to broaden their knowledge in the area of Machine Learning.

1.3 Structure

This book consists of the *Introduction* outlining its scope and purpose, five chapters and conclusion.

Chapter 2 *Understanding Machine Learning* presents the background and explains the logic behind machine learning process in conventional and quantum computing. It provides insight into those parts that are well documented and confirmed to be scientifically based and highlights the issues that are yet to be researched.

Chapter 3 *Defence* reviews the existing machine learning methods to ensure security and safety in cyber space for large scale networks, highly critical infrastructures, but also for personal use. It covers network- and host-based intrusion detection, software analysis, validation of secure configurations of systems, and hybrid defence solutions.

Chapter 4 *Attack* describes in detail the existing and potential applications of machine learning for cyber attacks. It elucidates selected methods for attack purposes and evolving threats, including from autonomous weapon systems.

Chapter 5 *International Resonance* outlines the ongoing debates around the AI integration, its high risks and offensive capabilities used for military purposes. It also reviews national and international efforts to develop and promote frameworks governing the use of AI for building peaceful, secure and stable digital environment.

Chapter 6 *Prospects* attempts to look into the future and scan the horizon of further developments and challenges up to 2050.

Conclusion completes the book and hints on the future work.

Reference

1. Russell SJ (2019) Human compatible: AI and the problem of control. Allen Lane, Penguin Books, p IX

Chapter 2
Understanding Machine Learning

> The mere formulation of a problem is often far more essential than its solution, which [...] requires creative imagination and marks real advances in science.
>
> *Albert Einstein, 1921*

Machine Learning is an emerging field of the Artificial Intelligence and data science. It is yet to be conceptualised and operationalised to be fully understood in its complexity and entirety. This chapter will consider it in a more detailed way through providing definitions to its constituting elements and analysing the learning process itself, and will review the accompanying factors around it.

When philosophers ask questions about knowledge, especially the fundamental, sceptical questions, they do not ask what it is that we know, but how, if at all, we can know what we know. This question is posed in a more practical form in Sir Tony Hoare's paper 'The Logic of Engineering Design.' Quite often computer programmers do not know how their own programmes work or how they will perform in certain circumstances. Sir Tony argues that this is a situation that can and should be avoided. His view is that if software engineering is carried out on a rigorous scientific basis, then programmers will have a way of knowing how their programmes will function and, more importantly, they will have a way of demonstrating that they will so function. Therefore, they can have demonstrable knowledge of what they know.

S. Abaimov and M. Martellini, *Machine Learning for Cyber Agents*,
Advanced Sciences and Technologies for Security Applications,
https://doi.org/10.1007/978-3-030-91585-8_2

Sir Tony argues that this knowledge can be gained by using the methods of propositional logic—that the conformity of an engineering design to a specification can, in principle, be established by a basic proof in propositional logic. His view is that computer science, as it matures, will come to rest on such perspicuous foundations.[1]

Philosophy of Engineering

The key to fully adopting machine learning as a technology is in understanding the processes behind it and, similar to the natural world, having a clear vision of why certain things work the way they work and avoiding false expectations.

2.1 Setting the Scene

Machine learning in its applied meaning is a technique used in the AI-systems development. AI is a very broad field, with multiple approaches, and *intelligence* itself has yet a variety of definitions. With reference to technical systems, their *intelligence* may be understood as "their ability to determine the most optimum course of action to achieve [...] goals in a wide range of environments."[2]

What is intelligence?

Q. What is artificial intelligence?

A. It is the science and engineering of making intelligent machines, especially intelligent computer programs. It is related to the similar task of using computers to understand human intelligence, but AI does not have to confine itself to methods that are biologically observable.

Q. Yes, but what is intelligence?

A. **Intelligence is the computational part of the ability to achieve goals in the world**. Varying kinds and degrees of intelligence occur in people, many animals and some machines.

Q. Isn't there a solid definition of intelligence that doesn't depend on relating it to human intelligence?

---

[1] Philosophy of Engineering, Volume 1 of the proceedings of a series of seminars, The Royal Academy of Engineering, 2010.

[2] 2018 UNDIR Report "The Weaponization of Increasingly Autonomous Technologies: Artificial Intelligence", as adapted from Shane Legg and Marcus Hutter, "A Collection of Definitions of Intelligence", Technical Report IDSIA-07–07, 15 June 2007, p. 9, last accessed on 16 March 2019 at http://www.unidir.ch/files/publications/pdfs/the-weaponization-of-increasingly-autonomous-technologies-artificial-intelligence-en-700.pdf.

A. Not yet. The problem is that we cannot yet characterize in general what kinds of computational procedures we want to call intelligent. We understand some of the mechanisms of intelligence and not others.[3]
John McCarthy (one of the founders of AI, the 1971 ACM A.M. Turing Award recipient)

For the purpose of this book, AI is defined as "the capacity of software to rely on and take decisions based on the provided knowledge rather than on the predefined algorithms" [1]. This capacity implies the software ability to first of all extract new knowledge from the available data, and secondly, depending on the level of autonomy, make decisions and even act, or activate some other system, to achieve specific goals.

Selected definitions of Artificial Intelligence

- "Artificial Intelligence … is the subfield of Computer Science devoted to developing programs that enable computers to display behaviour that can (broadly) be characterized as intelligent."[4]
- "At its simplest form, artificial intelligence is a field, which combines computer science and robust datasets, to enable problem-solving. It also encompasses sub-fields of machine learning and deep learning, which are frequently mentioned in conjunction with artificial intelligence. These disciplines are comprised of AI algorithms which seek to create expert systems which make predictions or classifications based on input data."[5]
- "Like computing, electrification, and the steam engine, AI is a general-purpose technology with transformative potential in a wide variety of sectors and with implications for the whole of society."[6]
- "Any artificial system that performs tasks under varying and unpredictable circumstances without significant human oversight, or that can learn from experience and improve performance when exposed to data sets.
- An artificial system developed in computer software, physical hardware, or other context that solves tasks requiring human-like perception, cognition, planning, learning, communication, or physical action.

[3] John McCarthy, What is Artificial Intelligence? 24 November 2004, p. 2, http://jmc.stanford.edu/articles/whatisai/whatisai.pdf.

[4] Richmond Thomason. Logic and artificial intelligence. In Edward N. Zalta, editor, The Stanford Encyclopedia of Philosophy. 2003. http://plato.stanford.edu/entries/logic-ai/.

[5] Artificial Intelligence (AI), last accessed 2021-09-01, https://www.ibm.com/cloud/learn/what-is-artificial-intelligence.

[6] OECD AI policy Observatory, https://www.oecd.ai/.

- An artificial system designed to think or act like a human, including cognitive architectures and neural networks.
- A set of techniques, including machine learning that is designed to approximate a cognitive task.
- An artificial system designed to act rationally, including an intelligent software agent or embodied robot that achieves goals using perception, planning, reasoning, learning, communicating, decision-making, and acting."[7]

The next term that has evolved and is currently in a wide use including in the latest AI Regulations[8] adopted by the European Union, is the "**AI systems**". However, from the perspective of a technical expert, calling the system an "AI system" may not be that straightforward, even if machine learning was used at certain stages in its development or decision-making process.

AI systems: selected definitions

Definition 1

"Artificial intelligence system" (**AI system**) means software that is developed with one or more of the techniques and approaches listed in Annex I[9] and can, for a given set of human-defined objectives, generate outputs such as content, predictions, recommendations, or decisions influencing the environments they interact with."[10] Annex 1 lists the following techniques and approaches:

(a) "Machine learning approaches, including supervised, unsupervised and reinforcement learning, using a wide variety of methods including deep learning;

[7] Daniel S. Hoadley, updated by Kelley M. Sayler, "Artificial Intelligence and National Security", 10 November 2020, Congressional Research Service Report, United States, pp. 1–2, https://fas.org/sgp/crs/natsec/R45178.pdf.

[8] Regulation of the European Parliament and of the Council Laying Down Harmonised Rules on Artificial Intelligence (Artificial Intelligence Act) and amending certain Union Legislative Acts, Brussels, 21.4.2021, at https://eur-lex.europa.eu/resource.html?uri=cellar:e0649735-a372-11eb-9585-01aa75ed71a1.0001.02/DOC_1&format=PDF.

[9] Annex 1 to Regulation of the European Parliament and of the Council Laying Down Harmonised Rules on Artificial Intelligence (Artificial Intelligence Act) and amending certain Union Legislative Acts, Brussels, 21.4.2021, at https://eur-lex.europa.eu/resource.html?uri=cellar:e0649735-a372-11eb-9585-01aa75ed71a1.0001.02/DOC_1&format=PDF.

[10] Regulation of the European Parliament and of the Council Laying Down Harmonised Rules on Artificial Intelligence (Artificial Intelligence Act) and amending certain Union Legislative Acts, Brussels, 21.4.2021, at https://eur-lex.europa.eu/resource.html?uri=cellar:e0649735-a372-11eb-9585-01aa75ed71a1.0001.02/DOC_1&format=PDF.

(b) Logic- and knowledge-based approaches, including knowledge representation, inductive (logic) programming, knowledge bases, inference and deductive engines, (symbolic) reasoning and expert systems;
(c) Statistical approaches, Bayesian estimation, search and optimization methods."

Definition 2

"**AI system** means a system that is either software-based or embedded in hardware devices, and that displays behaviour simulating intelligence by, inter alia, collecting and processing data, analysing and interpreting its environment, and by taking action, with some degree of autonomy, to achieve specific goals;

… An AI system is a machine-based system that can, for a given set of human-defined objectives, make predictions, recommendations, or decisions influencing real or virtual environments. AI systems are designed to operate with varying levels of autonomy."[11]

Definition 3

"**Al Systems** are information-processing technologies that embody models and algorithms that produce a capacity to learn and to perform cognitive tasks leading to outcomes such as prediction and decision-making in real and virtual environments.

Al Systems are designed to operate with some aspects of autonomy by means of knowledge modelling and representation and by exploiting data and calculating correlations. Al Systems may include several methods, such as but not limited to:

i. machine learning, including deep learning and reinforcement learning,
ii. machine reasoning, including planning, scheduling, knowledge representation and reasoning, search, and optimization, and
iii. cyber-physical Systems, including the Internet-of-Things, robotic Systems, social robotics, and human–computer interfaces which involve control, perception, the processing of data collected by sensors, and the operation of actuators in the environment in which Al Systems work."[12]

[11] Report on artificial intelligence: questions of interpretation and application of international law in so far as the EU is affected in the areas of civil and military uses and of state authority outside the scope of criminal justice (2020/2013(INI)). Committee on Legal Affairs Rapporteur: Gilles Lebreton, 4.1.2021, p. 6, https://www.europarl.europa.eu/doceo/document/A-9-2021-0001_EN.pdf.

[12] UNESCO AI ethical regulations https://unesdoc.unesco.org/ark:/48223/pf0000374266/PDF/374266eng.pdf.multi.

During more than half a century the AI-related research has passed through stages of the global interest and neglection, through inspiring expectations and disappointments that were partially related to the delay in the development of hardware technologies able to meet the AI needs. The AI research is usually interpreted as the study of **intelligent agents**, that may be represented by "any device that perceives its environment and takes actions that maximize its chance of successfully achieving its goals".[13]

The AI is categorized with reference to the human intelligence and capacity to perform tasks from the levels lower to beyond it, i.e. **narrow, general and super-intelligence**. *Artificial Narrow Intelligence* is a computer system capacity to implement specific tasks based on the analysis of the provided or generated data. *Artificial General Intelligence* is expected to be equal to the human intelligence, and *Artificial Superintelligence* will even surpass it.

Currently, the narrow AI has a wide application that is limited by risks stemming from the opacity of the algorithms decision-making, lack of predictability of the neural networks' behaviour, cyber vulnerabilities, errors and biases existing in the intelligent machines. The AI capacity to resolve issues at the level of complex human intelligence raises doubts and mistrust. The most visionary minds, though, keep predicting that the AI at a certain moment will match and outperform humans.[14] As per the forecasts, this moment of "singularity" may potentially happen somewhere from 20–50 years from now to even 120 years when AI can potentially implement all the jobs done by humans.[15,16] This will also allow fully autonomous system taking decisions independently and replacing human operators.

The DARPA *Perspective on AI* indicates the following capacities expected from the AI-systems: to perceive rich, complex and subtle information; learn within an environment; abstract to create new knowledge; reason to plan and to decide. DARPA proposed the intelligence scale through evaluation of *perceiving*, *learning*, *abstracting* and *reasoning* capacities.[17]

---

[13] Poole et al. [2].

[14] Kurzweil [3].

[15] V. C. Muller, N. Bostrom Future Progress in Artificial Intelligence: A Survey of Expert Opinion, 2013, at https://nickbostrom.com/papers/survey.pdf.

[16] K. Grace et al., When will AI exceed human performance? Evidence from AI experts, 3 May 2018, https://arxiv.org/abs/1705.08807.

[17] John Launchbury, A DARPA Perspective on Artificial Intelligence, https://www.darpa.mil/attachments/AIFull.pdf.

DARPA about AI development

DARPA highlights three historical periods in the AI development: *handicraft knowledge, statistical learning, contextual adaptation.*

During the **handicraft period**, the knowledge was codified through the computer acceptable language, with rules programmed by humans to resolve specific problems in a logical way (e.g., games, programmes for logistics, tax collection). The systems were good with reasoning, but unable to learn, to resolve unprogrammed issues, to deal with uncertainty or to act in the natural environment. These systems are no longer considered as AI, but they are still in use and demonstrate high performance, e.g. in automation. In 2016, the DARPA Cyber Grand Challenge demonstrated their strong advantages in cyber security.[18] Using the DARPA intelligence scale, they were good in reasoning, slightly satisfactory in perceiving, but zero in learning or abstracting.

The period defined by DARPA as **statistical learning,** creation of statistical models, allowed machines to process information, classify it and develop predictions. While perceiving and learning capacities have increased comparing to the first stage, reasoning however decreased. Abstracting capacity, generalization and application of knowledge in other areas, are demonstrated at the minimal level. Their reliability level is low, e.g. the chatbot Tay released by Microsoft in 2016 to engage in social conversation failed to cope with differentiation of the information content.[19]

DARPA associates the AI Third Wave with the **contextual adaptation** of the systems, their capacity to build explanatory models of the decisions taken and the environment. It is expected that at this stage, AI will be able to perceive, learn, reason at a high level, with at least satisfactory capacity to abstract and transfer knowledge. The developed models should be explainable, and their decision-making process understandable.[20]

The idea about thinking machines has been inspiring humanity for centuries. The **history** of Machine Learning is directly related to the AI history and dates back to the 1940s, when the first electronic computers were created and Turing conducted first experiments with machine intelligence.[21] It is considered that the official study of the computer-generated intelligence started in 1956 with the Dartmouth *Summer*

[18] Cyber Grand Challenge (CGC) (Archived), last accessed 2021-09-02, https://www.darpa.mil/program/cyber-grand-challenge.

[19] Oscar Schwartz, 2016, Microsoft's Racist Chatbot Revealed the Dangers of Online Conversation, 25 November 2019, IEEE Spectrum, https://spectrum.ieee.org/tech-talk/artificial-intelligence/machine-learning/in-2016-microsofts-racist-chatbot-revealed-the-dangers-of-online-conversation.

[20] John Launchbury, A DARPA Perspective on Artificial Intelligence, https://www.darpa.mil/attachments/AIFull.pdf.

[21] A. M. Turing, I. Computing machinery and intelligence. *Mind*, Volume LIX, Issue 236, October 1950, p. 454 https://doi.org/10.1093/mind/LIX.236.433; Norbert Wiener, The Human Use of Human Beings, Da Capo Press, 1950.

Workshop on Artificial Intelligence. This workshop launched a new scientific field and research in developing intelligent thinking machines able to learn. One of the earliest AI milestones was the computer checkers programme developed by one of the AI pioneers and the IBM engineer, Arthur Samuels. While creating a computer programme able to play checkers, instead of programming, Samuels provided the computer with information about the rules of the game. It was tasked to play checkers with itself, finally winning a victory over its creator.[22]

Discussions on machine learning capacities inspired most brilliant visionary minds, and it was by the end of 1950s when machine learning was shaped as a separate term and scientific direction.[23,24] The collection of numerous papers on AI research, the first AI anthology "Computers and Thought", was published in 1963.[25] A series of academic publications followed based on extensive research in machine learning for patterns recognition and classification.[26,27,28] Artificial neural networks, inspired by a human brain, were developed and experiments with machine learning were conducted using symbolic methods, statistical models,[29] probabilistic reasoning, perceptrons. The term neural network was already in use, though reflecting an initial stage of learning. However, as the innovative probabilistic approaches required more efforts in generating data and higher hardware costs,[30] while the logical knowledge-based systems were easier, more practical and cheaper in use, the preferential balance was in favour of the expert systems.[31] Thus, during a decade they were dominating the research area and practical application, and the non-linear, flexible and adaptive algorithms and neural networks were nearly ignored. With the Internet coming into life, the situation changed and the learning systems received due attention.[32]

The difference between the programming and machine learning approaches can be illustrated through a simplified scheme presented in Fig. 2.1.

---

[22] IBM official web-site, The IBM 700 Series Computing Comes to Business, https://www.ibm.com/ibm/history/ibm100/us/en/icons/ibm700series/impacts/#:~:text=Playing%20checkers%20on%20the%20701,on%20an%20IBM%207094%20computer.

[23] Samuel Arthur (1959). "Some Studies in Machine Learning Using the Game of Checkers". *IBM Journal of Research and Development*. 3(3): 210–229. CiteSeerX 10.1.1.368.2254. https://doi.org/10.1147/rd.33.0210.

[24] R. Kohavi and F. Provost, "Glossary of terms," Machine Learning, vol. 30, no. 2–3, pp. 271–274, 1998.

[25] Feigenbaum and Feldman [4].

[26] Nilsson [5].

[27] Duda and Hart [6].

[28] S. Bozinovski "Teaching space: A representation concept for adaptive pattern classification" COINS Technical Report No. 81–28, Computer and Information Science Department, University of Massachusetts at Amherst, MA, 1981. https://web.cs.umass.edu/publication/docs/1981/UM-CS-1981-028.pdf.

[29] Sarle Warren (1994). "Neural Networks and statistical models". CiteSeerX 10.1.1.27.699.

[30] Stuart and Peter [7], p. 488. ISBN 978-0137903955.

[31] Langley Pat (2011). "The changing science of machine learning". Machine Learning. 82(3): 275–279. https://doi.org/10.1007/s10994-011-5242-y.

[32] Nilsson and Kaufman [8], Poole et al. [2].

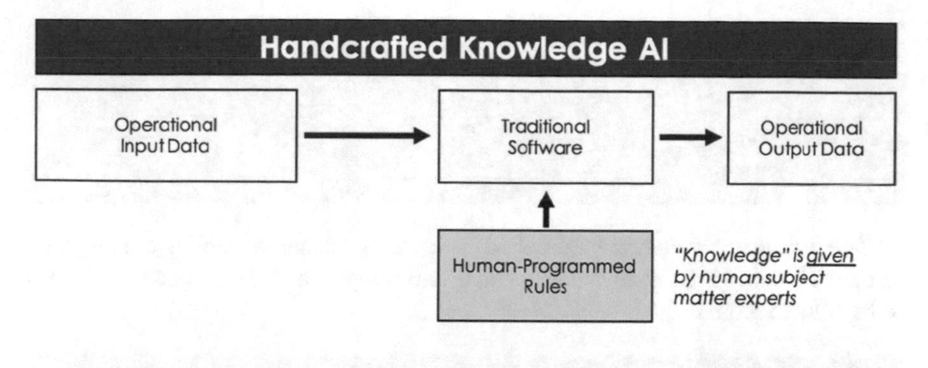

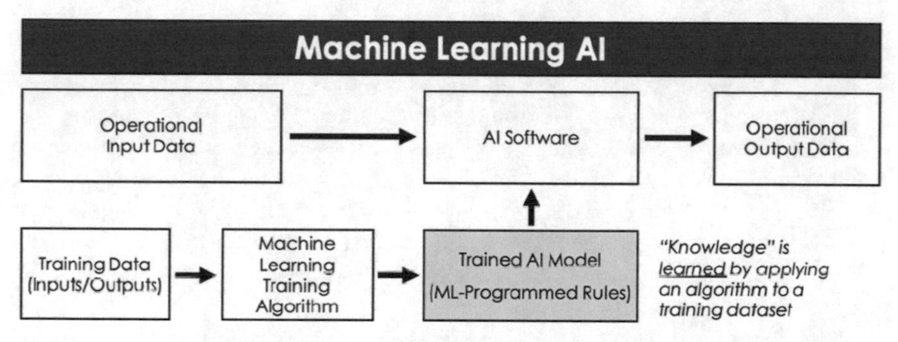

Fig. 2.1 Approaches to AI development. *Source* G. Allen [33]

Machine Learning strongly defined itself as a field of computer sciences, and a cross cutting area with the data science in the 1990s. Models built on statistical analysis and probability theory outweighed the symbolic approaches used in AI. Machine learning offered advanced solutions in data procession and practical application of the learning results.

How is the AI research done?

AI research has both theoretical and experimental sides. The experimental side has both basic and applied aspects. There are two main lines of research. One is biological, based on the idea that since humans are intelligent, AI should study humans and imitate their psychology or physiology. The other is phenomenal, based on studying and formalizing common sense facts about the world and the problems that the world presents to the achievement of goals. The two

[33] Greg Allen, Understanding AI Technology: A concise, practical, and readable overview of Artificial Intelligence and Machine Learning technology designed for non-technical managers, officers, and executives, Joint Artificial Intelligence Center, April 2020, p. 8, https://www.ai.mil/docs/Understanding%20AI%20Technology.pdf.

approaches interact to some extent, and both should eventually succeed. It is a race, but both racers seem to be walking.[34]
John McCarthy (one of the founders of AI, the ACM A.M. 1971 Turing Award recipient)

The discussion about the hard-coded logic versus machine learning are ongoing, as both of them have the advantages and disadvantages, and the potential that is yet to be fully explored.

AI paradigms

There are two quite different paradigms for AI. Put simply, the **logic-inspired paradigm** views sequential reasoning as the essence of intelligence and aims to implement reasoning in computers using hand-designed rules of inference that operate on hand-designed symbolic expressions that formalize knowledge.

The **brain-inspired paradigm** views learning representations from data as the essence of intelligence and aims to implement learning by hand-designing or evolving rules for modifying the connection strengths in simulated networks of artificial neurons.

In the logic-inspired paradigm, a symbol has no meaningful internal structure: its meaning resides in its relationships to other symbols which can be represented by a set of symbolic expressions or by a relational graph. By contrast, in the brain-inspired paradigm the external symbols that are used for communication are converted into internal vectors of neural activity and these vectors have a rich similarity structure. [...].

The main advantage of using vectors of neural activity to represent concepts and weight matrices to capture relationships between concepts is that this leads to automatic generalization. If Tuesday and Thursday are represented by very similar vectors, they will have very similar causal effects on other vectors of neural activity. This facilitates analogical reasoning and suggests that immediate, intuitive analogical reasoning is our primary mode of reasoning, with logical sequential reasoning being a much later development.[35]

Yoshua Bengio, Yann Lecun, Geoffrey Hinton (the 2018 ACM A.M. Turing Awards recipients for Deep Learning)

Computers undoubtfully can do some things better than people and are outperforming humans in some specific tasks. Deep Learning algorithms have no limits in

[34] John McCarthy, What is Artificial Intelligence? 24 November 2004, p. 12, http://jmc.stanford.edu/articles/whatisai/whatisai.pdf.

[35] Yoshua Bengio, Yann Lecun, Geoffrey Hinton. "Turing lecture: Deep Learning for AI", Communications of the ACM, July 2021, Vol. 64, No. 7, pp. 58–65, 10.1145/3448250 at https://cacm.acm.org/magazines/2021/7/253464-deep-learning-for-ai/fulltext.

memory and benefit from larger amounts of data. They can see, listen, read, write, translate from languages simultaneously voicing the translated text, transcribing it and generating novel texts. To challenge the researchers and share findings, the **AI competitions** are organized globally.

AI competitions

"The **International Aerial Robotics Competition** (IARC),[36] has been organized for about 30 years for students challenging them with the missions impossible. Its aim is to advance the research and improve the performance of aerial robots.

In 2004, DARPA recognized that none of its cars were able to cross a desert in a self-driving mode due to lack of vision. While in 2005, already five cars were able to successfully demonstrate their self-driving capacities through 132 miles of the **DARPA Autonomous Vehicle Grand Challenge**. The prize was 2 million US dollars.[37] It was followed by the **DARPA Urban Challenge** for self-driving cars challenged by the urban traffic, and later by the **Multi Autonomous Ground-robotic International Challenge** with a simulated military exercise (prize 1,6 million US dollars).

In 2004–2006, the **Face Recognition Grand Challenge** tested the face recognition techniques aiming "to promote and advance face recognition technology designed to support existing face recognition."[38] though having been organized for only two years, it "developed new face recognition techniques and prototype systems while increasing performance by an order of magnitude,… and improved the capabilities of automatic face recognition systems through experimentation with clearly stated goals and challenge problems".

In 2007, the **Pittsburgh Brain Activity Interpretation Competition** offered $22,000 for the prediction of "what individuals perceive and how they act and feel in a novel Virtual Reality world involving searching for and collecting objects, interpreting changing instructions, and avoiding a threatening dog" based on the fMRI data. A year before, the same competition "involved prediction of the subjective experience of movie viewing from fMRI" and gathered participants from 31 countries.[39]

[36] International Aerial Robotics Competition, IARC, http://www.aerialroboticscompetition.org/.

[37] The Grand Challenge, DARPA, https://www.darpa.mil/about-us/timeline/-grand-challenge-for-autonomous-vehicles.

[38] Overview of FRGC, https://web.archive.org/web/20080410072057/http://face.nist.gov/frgc/.

[39] The Pittsburgh Brain Activity Interpretation Competition, https://web.archive.org/web/20080310070925/http://www.ebc.pitt.edu/PBAIC.html.

In 2008, the **American Meteorological Society** organized an AI competition "to use the supplied data sets to classify the observed precipitation type at the ground into one of three categories: liquid, frozen, or none." The data was used "from meteorological analyses of environmental conditions close in time and space to the actual observations, along with data from a polarimetric radar"[40]

The **British Computer Society Machine Intelligence competition**[41] has been organised by the Specialist Group on Artificial Intelligence and is based on a short demonstration of the AI-enhance software or hardware. The prize is £500.

Battlecode, an MIT-hosted programming competition, is a unique challenge that combines battle strategy, software engineering, and artificial intelligence. Student teams aim to write the best player program for the Battlecode real-time strategy game.[42]

Since 2015, **Roborace**, "the world's first extreme competition of teams developing self-driving AI",[43] has been demonstrating the world's fastest autonomous racing cars competing in an extreme environment.

In 2017, IBM initiated the AI competition **AI for Good** to inspire researchers and developers to demonstrate creativity in resolving societal issues of the global dimension. "Each team's technologies are evaluated across four dimensions: achieved technical impact, evidenced real-world impact, scalability of real-world impact, and ethics and safety."[44]

Machine learning advancements were fuelled by the increased computational power, availability of larger volumes of data samples, open source programming libraries, decreased costs of hardware processing power and enlarged storage capacities, optimisation of approaches.

Machine Learning Accelerators

To improve the training performance, instead of Central Processing Units (CPUs) the following processing chips are used:

- GPU—Graphics processing unit
- FPGA—Field-programmable gate array

[40] 2008 AMS Artificial Intelligence Competition, https://web.archive.org/web/20091113074224/http://www.nssl.noaa.gov/ai2008/.

[41] British Computer Society Machine Intelligence Competition, http://www.bcs-sgai.org/micomp/intro.php.

[42] What is Battlecode?, MIT Admissions, https://mitadmissions.org/help/faq/battle-code/.

[43] Roborace, https://roborace.com/.

[44] XPrize AI, https://www.xprize.org/prizes/artificial-intelligence.

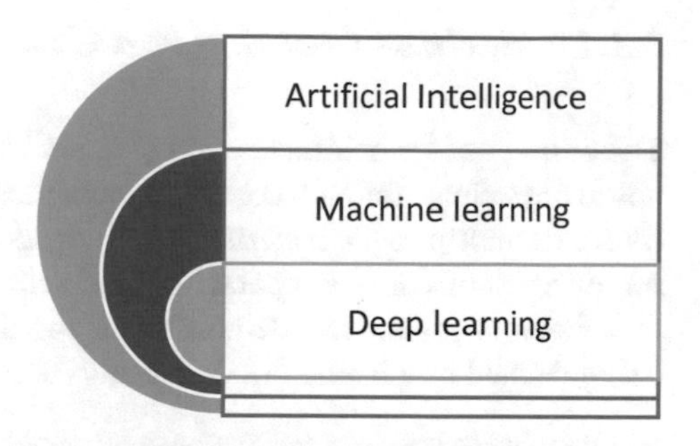

Fig. 2.2 Correlation between deep learning, machine learning and artificial intelligence as a science field

- TPU—Tensor processing unit
- ASIC—Application-specific integrated circuit.

Of note: Smartphone Google Pixel 6 is reported to use TPU.[45]

The Central Processing Units (CPUs) were replaced by Graphics Processing Units (GPUs), offering an order of magnitude in faster learning; and the cloud technologies allowed extensive storage capacities and accessible computational power to everyone. In the first decade of the twenty-first century, the neural networks development got a new impulse through all these technological achievements, and multilayered Deep Neural Networks have become more in use being followed by the revolutionary wide application of deep learning (see Fig. 2.2). Described only in a science fiction previously, speech and image recognition, machine translation, autonomous systems became a reality.

The new technologies challenged human intelligence by their own way of "thinking". Lack of clarity in understanding the core machine learning processes with the application of the neural networks and deep learning, a so-called "black box" problem, lack of predictability in decisions, raised strong reliability, safety and security concerns. The research has moved to developing the explainable AI (XAI),[46] trustworthy algorithms able to "explain" the decision-making process.

2.2 Conceptual and Operational Landscape

> The tension between reasoning and learning has a long history, reaching back at least as far as Aristotle.[47]
>
> *Leslie Gabriel Valiant (the 2010 ACM A.M. Turing Award recipient)*

[45] Google Tensor debuts on the new Pixel 6 this fall, Google, 2 August 2021, https://blog.google/products/pixel/google-tensor-debuts-new-pixel-6-fall/.

[46] Explainable Artificial Intelligence (XAI), DARPA, https://www.darpa.mil/program/explainable-artificial-intelligence.

[47] Valiant [9], Chap. 7, NY 2013, e.

2.2.1 Machine Learning as a Concept

The concept of "**machine learning**" is not new, and yet, its definitions vary and are not always clear. Being a complex phenomenon and depending on the approaches, it reveals multiples characteristics and capacities. The discussion on whether it is an AI, or a part of AI, or a separate field of science are still ongoing.

Literature review has demonstrated the diversity in this term perception, from being defined as a method of data analysis, to being a science.

Machine Learning defined

"Machine learning is **a method of data analysis** that automates analytical model building. It is **a branch of artificial intelligence** based on the idea that systems can learn from data, identify patterns and make decisions with minimal human intervention."[48]

"Machine learning is a **method of data analysis** that automates analytical model building. It is a branch of artificial intelligence based on the idea that systems can learn from data, identify patterns and make decisions with minimal human intervention…. machine learning is a specific subset of AI that trains a machine how to learn."[49]

"Machine learning is **the study of computer algorithms** that improve automatically through experience and by the use of data."[50]

"Machine learning is the concept that a computer program can learn and adapt to new data without human intervention. Machine learning is **a field of artificial intelligence** that keeps a computer's built-in algorithms current regardless of changes in the worldwide economy."[51]

"Machine learning is a **subfield of artificial intelligence**, which is broadly defined as the capability of a machine to imitate intelligent human behaviour. Artificial intelligence systems are used to perform complex tasks in a way that is similar to how humans solve problems."[52]

"Machine learning is the **science** of getting computers to act without being explicitly programmed".[53]

---

[48] SAS, analytical agency, https://www.sas.com/en_gb/insights/analytics/machine-learning.html#:~:text=Machine%20learning%20is%20a%20method,decisions%20with%20minimal%20human%20intervention.

[49] Machine Learning, SAS, https://www.sas.com/en_us/insights/analytics/machine-learning.html.

[50] Mitchell [10], ISBN 0-07-042807-7. OCLC 36417892.

[51] Machine learning, Investopedia, https://www.investopedia.com/terms/m/machine-learning.asp.

[52] Machine learning, explained, Sloan School, MIT, https://mitsloan.mit.edu/ideas-made-to-matter/machine-learning-explained.

[53] Machine learning, Coursera, https://www.coursera.org/learn/machine-learning.

"[M]achine learning ... is the **science** of creating intelligent computer programs that can automatically improve their performance through experience (i.e., 'learning').[54]

For the purpose of this book, as a scientific branch, Machine Learning is defined a subfield of computer sciences, aimed at studying the structure of data and fitting that data into mathematical models that can be interpreted and further used by other programs or human operators. Simultaneously, as a method, machine learning is one of techniques applied in software development.

2.2.2 *Algorithms and Their Application*

The concept of the "**machine learning algorithm**" also requires a more precise definition.

Algorithm is defined as "a set of mathematical instructions that must be followed in a fixed order, and that, especially if given to a computer, will help to calculate an answer to a mathematical problem."[55] It is a programmed guide explaining to the computer system how to resolve this or that problem.

Even something as simple as a program that saves previously entered commands into a file can be called a machine learning algorithm.[56] This being due to the fact that it "remembers" selected actions, and can suggest the most used commands so that the user does not need to type the entire string of symbols. This type of algorithm is currently used in many smartphone applications and terminal emulators. A very similar algorithm is used to predict and suggest the next word in search engines and recent text editing applications. It is worth noting that any machine learning algorithm may be of dual use, being applied with equal efficiency both in cyber defence and offence.

---

[54] Jacopo Bellasio, Erik Silfversten, "The Impact of New and Emerging Technologies on the Cyber Threat Landscape and Their Implications for NATO", in Cyber Threats and NATO 2030: Horizon Scanning and Analysis', ed. by A. Ertan, K. Floyd, P. Pernik, Tim Stevens, CCDCOE, NATO Cooperative Cyber Defence Centre of Excellence, King's College London, Williams and Mary, 2020, p. 90 at https://ccdcoe.org/uploads/2020/12/Cyber-Threats-and-NATO-2030_Horizon-Scanning-and-Analysis.pdf.

[55] Algorithm, Cambridge dictionary, https://dictionary.cambridge.org/dictionary/english/algorithm.

[56] *For more detail, see* Das, K., & Behera, R. N. (2017). A survey on machine learning: concept, algorithms and applications. International Journal of Innovative Research in Computer and Communication Engineering, 5(2), 1301–1309.

Neural network with 1 neuron:

$$f(x) = a \cdot x$$

Neural network with 2 neurons:

$$f(x) = a \cdot x_1 + b \cdot x_2$$

Neural network with *n* neurons:

$$f(x) = w_0 + K \cdot \sum_{i=1}^{n} w_i x_i$$

a, b, w – weights
K – number of neurons

Fig. 2.3 A simplified machine learning equation

Machine learning algorithm

A computer program is said to learn from experience *E* with respect to some class of tasks *T* and performance measure *P* if its performance at tasks in *T*, as measured by *P*, improves with experience *E*.[57]

Tom Mitchel, 1997

There is a mathematical function inside every machine learning model, with parameters and weights that get adjusted during the training process. Figure 2.3 presents the simplified machine learning equation, based on an artificial neural network with a single layer.

The key question with relation to algorithms would be the identification of their **minimal but sufficient constituting elements**. Machine learning algorithm can be a simple history file, as described above, but usually it is understood as a random tree, a neural network, a support vector machine, or any kind of model based on a mathematical method, that can approximate a function using sample data.

What is important for AI is to have algorithms as capable as people at solving problems. The identification of subdomains for which good algorithms exist is important, but a lot of AI problem solvers are not associated with readily identified subdomains.[58]

John McCarthy (one of the founders of AI, the 1971 ACM A.M. Turing Award recipient)

Machine learning can help in the development of IDS that can detect zero-day exploits or predict early signs of a cyber attack and deploy countermeasures before the attackers can reach critical systems inside the network, such as network access storage or industrial controls.

[57] Mitchell [10], p. 2. ISBN 978-0-07-042807-2.

[58] John McCarthy, What is Artificial Intelligence? 24 November 2004, p. 7, http://jmc.stanford.edu/articles/whatisai/whatisai.pdf.

The most widely spread machine learning algorithms are artificial neural networks (ANNs) that were designed with the idea to copy the interaction of the human neurons. The model developed by Hebb[59] helps to explain the relations between artificial neurons (nodes) that "strengthen" if the two neurons are activated at the same time and weakens if they are activated separately. The word "weight" is used to describe these relationships, and neurons tending to be both positive or both negative are described as having strong positive weights. Those neurons tending to have opposite weights develop strong negative weights. The ANNs have a complex structure and are able to implement more complex tasks, e.g. identify patters in large data.

Hebb's model of the brain cells' interaction
When one cell repeatedly assists in firing another, the axon of the first cell develops synaptic knobs (or enlarges them if they already exist) in contact with the soma of the second cell.[60]

Donald Hebb, 1949

A decade ago, only people who had a full understanding of machine learning and deep learning algorithms could use them in their applications. Currently, integrating a pre-trained neural network into an application is a lot simpler and a lot more straightforward. Machine learning inventions are already broadly integrated in industries, research, and ordinary life. For instance, Programming libraries like PyTorch and Tensorflow have tools that enable machine learning engineers to publish pretrained neural networks on GitHub, GitLab, or any other repository, and make them accessible to other developers. Basic machine-learning-based applications can be deployed with minimal knowledge about the underpinning mathematics of the neural networks; and multiple applications can be deployed on the same platform to form a full end-to-end AI system. Being used as an optimisation technique, it also improved the performance of the existing systems, directly, by using machine learning in the decision-making process, and indirectly, by using it to calculate the optimal design of a newly developed system.

Machine learning algorithms demonstrate high accuracy of decision making in complex tasks, especially in those that cannot be represented in a simple way suitable for conventional computational methods. Their applications can range from video surveillance to the network intrusion detection of enterprise networks. Under the human supervision, the *narrow AI* performs a wide range of supporting functions. It collects, analyses and stores data, plans, monitors performance and evaluates the outcomes, communicates with humans and recognizes objects. It uses the latest achievements of modern sciences—mathematics, biology, engineering, building its action on logics and symbols, stored or searched knowledge, adaptive mechanisms for orientation in environment, neural networks, probabilistic methods, morphological

[59] Hebb [11].
[60] *ibid.*

computation.[61] Typically, those applications and methods are deployed as a subprocess in a system to solve specific real-life problems, such as a process of object recognition as a part of self-driving vehicles, natural language processing as a part of voice assistant in a smart phone, and numeric pattern recognition in network intrusion detection. Neural networks are currently applied in specialised software used in banking, business analytics, education, statistics, and most advanced areas of health care. For example, in oncology to mimic the progression, growing and infiltration of the Pancreatic and Glioma cancer cells into the healthy biological tissues.[62] The use of machine learning in **identifying cyber-attacks** is actively studied in the academic community and presents a promising area of research with an enormous defence potential.

2.2.3 Models

Machine learning **model** is created during the training stage, and in simple words, may be defined as an **operationalised algorithm**. The training process is automated, but each task is unique and requires human supervision to confirm and verify the quality of the model.

> Machine Learning systems are different in that their "knowledge" is not programmed by humans. Rather, their knowledge is learned from data: a Machine Learning algorithm runs on a training dataset and produces an AI model. To a large extent, Machine Learning systems program themselves. Even so, humans are still critical in guiding this learning process. Humans choose algorithms, format data, set learning parameters, and troubleshoot problems.[63]
>
> *Greg Allen, Understanding AI Technology, 2020*

Models are trained using large datasets before the initial deployment, adjusted and optimised to have the highest accuracy and precision for a particular implementation or task. The initial training requires high computational resources, but a lot less computing power once the model is already trained. Pre-trained models can be deployed in less powerful devices and be available to a larger number of

---

[61] Cangelosi and Fischer [12], https://www.researchgate.net/publication/283812826_Embodied_Intelligence.

[62] Barbara Kenner et al., Artificial Intelligence and Early Detection of Pancreatic Cancer, Pancreas, Volume 50, Issue 3, March 2021, pp. 251–279, https://doi.org/10.1097/MPA.0000000000001762 (https://journals.lww.com/pancreasjournal/Fulltext/2021/03000/Artificial_Intelligence_and_Early_Detection_of.1.aspx).

[63] Greg Allen, Understanding AI Technology: A concise, practical, and readable overview of Artificial Intelligence and Machine Learning technology designed for non-technical managers, officers, and executives, Joint Artificial Intelligence Center, April 2020, p. 3, https://www.ai.mil/docs/Understanding%20AI%20Technology.pdf.

users, including through open-source libraries, e.g. Tensorflow,[64] PyTorch,[65] "Neural Network Libraries",[66] and MXNet.[67]

When acting in a predictable environment, conventional systems can fully rely on the linear and static models. Machine learning enables orientation, decision-making and action in the complex environment through collecting information, its analysis and subsequent adjustments. A set of joint **models** may result in partial or even full autonomy, independent decision-making process based on the series of mathematically defined algorithms. The task itself may not necessarily be formulated, and this is for the machine to analyse the incoming signals or provided data set, identify the correlations and patterns, classify them and adjust the learning algorithms to come up with the most optimum solution and operational model in the unknown complex environment. Currently, the pattern recognition (voice, image, target, intrusion detection, etc.) is based on this method.

Machine learning models are aimed to be cross-platform so they could be potentially deployed on any system. However, the prototype and proof-of-concept systems, that exist today, are constrained to specific versions of specific hardware, software, and programming libraries. Models have their limited **life cycle**, as their continuous learning is a labour intense and expensive process. Systems that continue learning even after the deployment, have to be closely monitored for errors and defects that might be introduced over time.

Model explainability is one of the biggest challenges in machine learning today, and the non-transparent way some models take decisions is defined as the interpretability problem. Machine learning is normally accompanied by large volumes of data, and full transparency of training may not be achievable. It is especially worrying that certain "black box" models, such as deep neural networks, are deployed to production and are running critical systems, such as security cameras, smartphone, etc. Not even the developers of these algorithms always understand why exactly the algorithms make the decisions they do—or even worse, how to prevent an adversary from exploiting them.

2.2.4 Methods

Machine learning typically implies that a computer program can extract, record and adapt to the new data without human intervention. This process can be implemented through multiple ways and based on the task at hand and the available data samples there may be selected one method or another.[68]

[64] Tensorflow, https://www.tensorflow.org/.

[65] PyTorch, https://pytorch.org/.

[66] Neural Network Libraries by Sony, https://nnabla.org.

[67] Apache MXNet, library for deep learning, http://mxnet.incubator.apache.org/index.html.

[68] Buskirk, T. D., Kirchner, A., Eck, A., & Signorino, C. S. (2018). An introduction to machine learning methods for survey researchers. Survey Practice, 11(1), 2718.

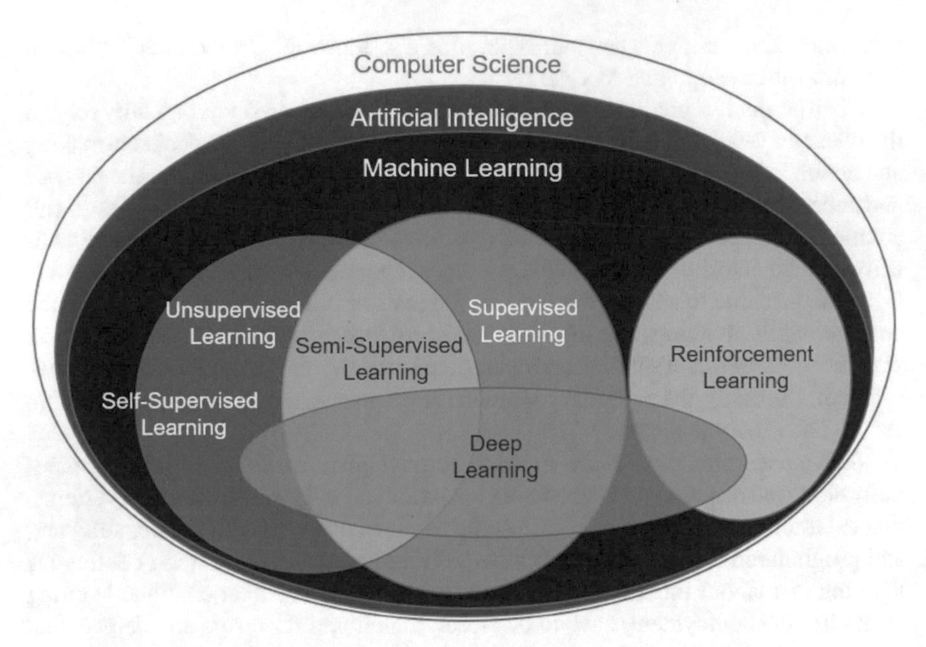

Fig. 2.4 Methods of machine learning

Machine learning can be **supervised, semi-supervised, and unsupervised**. There also exist a **reinforcement learning**. The difference between them is important in understanding the AI functionality and its use. In *supervised learning*, the input patterns are provided with the desired outcomes. Programs learn from provided patterns and compare them with the expected outputs. In *semi-supervised learning*, a model is pre-trained with a supervised method (labelled data), and further used for unsupervised learning (unlabelled data). In *unsupervised learning*, the machine must classify raw data into types (labels) to optimize the probability of generating the best possible solution without provided desired outcomes. Machines learn from the unlabelled data and self-identify patterns in it. This learning is perceived to be most suitable for a system to orient in the unknown environment, also to identify unknown threats in the traffic flow. The *reinforcement learning* is done through awards received for each correctly resolved challenge.

The **deep learning** technique, based on using multi-layered deep neural networks, can be applied through all machine learning approaches. Though it is the most challenging method and least predictable, it is considered that "[a]s of 2020, deep learning has become the dominant approach for much ongoing work in the field of machine learning".[69] (Fig. 2.4).

Even though machine learning is a rising and promising field, none of the methods can fully guarantee a perfect real-life accuracy in performance for the underlying technology and cannot be fully trusted with the final decision.

---

[69] Alpaydin [13], pp. xix, 1–3, 13–18.

Cases of AI failure

2010s: racial and other bias in the use of AI for criminal sentencing decisions and findings of creditworthiness.
2016: Tay AI, deployed by Microsoft, failed to differentiate the contextual information and twitted using abusive language.
2016: Deepfake fuelled fake news.
2018: Accessible photo face filters gave a new tool to crime, specifically dating fraud.
2019: A facial-recognition system identified three-time Super Bowl champion Duron Harmon of the New England Patriots, Boston Bruins forward Brad Marchand, and 25 other New England professional athletes as criminals. Amazon's Rekognition software falsely matched the athletes to a database of mugshots in a test organized by the Massachusetts chapter of the American Civil Liberties Union (ACLU). Nearly one-in-six athletes were misidentified. "This technology is flawed."[70]
2020: AI-Powered 'Genderify' platform shut down after bias-based backlash. Starsky Robotics, the unmanned vehicles start up, failed. Uber "walked away" from AI and sold the driverless vehicles division. Facebook AI Blenderbot lost to Pandorabot Kuki at the "first date" failing to make sense in the words used, e.g. "It is exciting that I get to kill people."[71]

Academically speaking, a method can be trusted when it can be predicted, or at least explained. Currently, the machine learning is neither fully understood, nor is a predictable technology, especially when it comes to the unsupervised learning. The practical implementation of the inspiring idea to copy the work of the human neurons, proved that in fact they are too complex to be represented in an equation or follow a simple algorithm. Human mind goes far beyond memorisation and pattern recognition. Furthermore, many modern approaches aim to use metadata—an "envelope" with values describing traffic flow and volume, etc., instead of data, which takes an evolutionary path, different from human intelligence.

The existing machine learning methods allow programs to learn, adjust and adapt. However, there are still many unknowns in the core machine learning process. For example, it is not yet explainable how exactly the decision is made by non-linear algorithms, a so-called "black box" decision-making process.

[70] Facial recognition technology falsely identifies famous athletes, 21 October 2019, https://www.aclum.org/en/news/facial-recognition-technology-falsely-identifies-famous-athletes/#athletes.

[71] 2020 in Review, 10 AI Failures, https://syncedreview.com/2021/01/01/2020-in-review-10-ai-failures/.

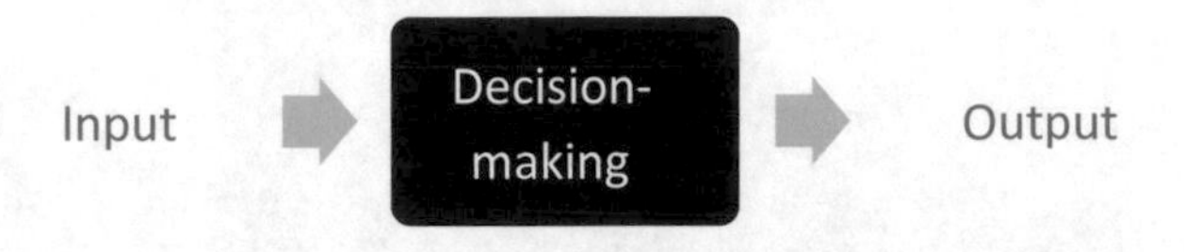

Fig. 2.5 The machine learning process presented as a "Black box"

> An important feature of a learning machine is that its teacher will often be very largely ignorant of quite what is going on inside, although he may still be able to some extent to predict his pupil's behaviour.... Most of the programmes which we can put into the machine will result in its doing something that we cannot make sense of at all, or which we regard as completely random behaviour.[72]
>
> *Alan Turing, computer and AI pioneer, 1950*

Black box technology by definition cannot be trusted and, in an ideal scenario, the AI-system has to be fully explained and predictable before being used. The issue of explainability requires a wider coverage and will be presented in more detail in the following subchapters.

2.3 Explainability of Machine Learning

For a better understanding of the machine learning process, it is essential to consider the work of the AI models. Their predictability levels differ, with the most problematic ones being the neural networks, that though ensuring high accuracy of decisions taken, have a very low level of explainability as compared to other algorithms (e.g., decision trees).

The machine learning-based models, especially in deep learning, implement the learning process in a covert way, and many do not allow to fully understand their function or the logic behind it. Simply put, there is data on the input, and there is data on the output, while **what happens in between is unknown and cannot be represented as an equation or a logical flowchart** (see Fig. 2.5).

Mathematics behind this method is known, and it is possible to extract weights and bias from the already trained model of a neural network. The decisions are made with a certain level of certainty and accuracy. Yet, it is still unknown why machine learning models behave the way they behave. This is considered to be one of the biggest challenges in the application of the AI techniques. It makes the machine decisions non-transparent and often incomprehensible even to the experts or the developers themselves, reducing trust in the use of this approach and in the AI in general.

The complexity of the neurons connections in deep neural networks is demonstrated in Fig. 2.6.

[72] A. M. Turing, I. Computing machinery and intelligence. *Mind*, Volume LIX, Issue 236, October 1950, p. 458. https://doi.org/10.1093/mind/LIX.236.433.

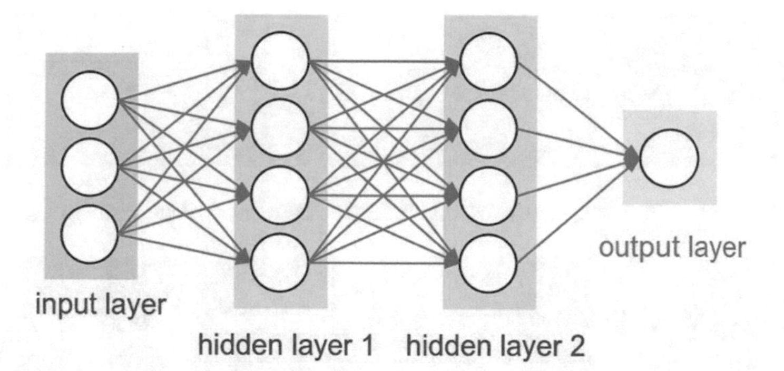

Fig. 2.6 Layers in a deep learning process

A neural network has three types of layers: input, hidden and output layers. A normal neural network would usually have no more than two hidden layers and linear functions. The neural networks with more than three hidden layers are considered as "deep" and their application is called **deep learning.** It came into use in the first decade of 2000s, making *a paradigm shift* allowing for the major AI-systems breakthroughs and winning the major AI competitions, e.g. enhanced computer vision, speech recognition, understanding complex sentences, medical diagnostics, etc.

Deep learning, the only technique used for supervised, semi-supervised, unsupervised, and reinforcement learning at the same time, has already asserted itself as one of the most versatile techniques with feasible advantages over the others. These advantages include the speed of procession of big amounts of data, enhanced learning of feature representations, reducing time and needs in manual feature identification, adaptability to tasks. The main areas of research in deep learning are the architecture of the networks and learning optimisation. The open-source programming libraries contribute to the extensive research in the deep learning application in intrusion detection systems. However, as currently it is impossible to explain with certainty what happens in the hidden layers when the input is processed into the output decision, both the neural networks and the deep learning methods itself, are characterized by the **lowest level of explainability, obscurity in decision-making and predictions**.

Difference between "classical" machine learning and deep learning

The way in which deep learning and machine learning differ is in how each algorithm learns. Deep learning automates much of the feature extraction piece of the process, eliminating some of the manual human intervention required and enabling the use of larger data sets. You can think of deep learning as "scalable machine learning" [....] Classical, or "non-deep", machine learning is more dependent on human intervention to learn. Human experts determine the hierarchy of features to understand the differences between data inputs, usually requiring more structured data to learn.[73]

IBM

As any technique, deep learning has its advantages and disadvantages.

Deep Learning: Pros

Research on artificial neural networks was motivated by the observation that human intelligence emerges from highly parallel networks of relatively simple, non-linear neurons that learn by adjusting the strengths of their connections. This observation leads to a central computational question: How is it possible for networks of this general kind to learn the complicated internal representations that are required for difficult tasks such as recognizing objects or understanding language? Deep learning seeks to answer this question by using many layers of activity vectors as representations and learning the connection strengths that give rise to these vectors by following the stochastic gradient of an objective function that measures how well the network is performing. [...] We believe that **deep networks excel** because they exploit a particular form of compositionality in which features in one layer are combined in many different ways to create more abstract features in the next layer.[74]

Yoshua Bengio, Yann Lecun, Geoffrey Hinton (the 2018 Turing Awards recipients for Deep Learning), 2021

Deep Learning: Cons

Despite the impressive results, deep learning has been criticised for brittleness (being susceptible to adversarial attacks), lack of explainability (not having a formally defined computational semantics or even intuitive explanation, leading to questions around the trustworthiness of AI systems), and lack of

[73] What is Artificial Intelligence, IBM, https://www.ibm.com/cloud/learn/what-is-artificial-intelligence.

[74] Yoshua Bengio, Yann Lecun, Geoffrey Hinton, "Turing lecture: Deep Learning for AI", *Communications of the ACM*, July 2021, Vol. 64 No. 7, pp. 58–65 10.1145/3448250 at https://cacm.acm.org/magazines/2021/7/253464-deep-learning-for-ai/fulltext.

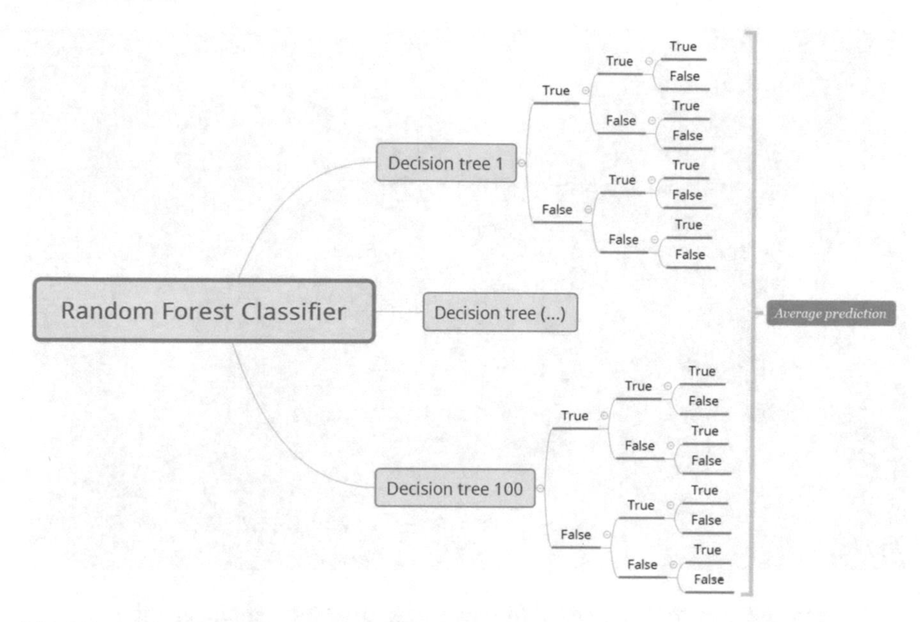

Fig. 2.7 *Random forest* graphic presentation

parsimony (requiring far too much data, computational power at training time or unacceptable levels of energy consumption)[75]
Artur D'Avila Garcez, Luis C. Lamb, Neurosymbolic AI: The 3rd Wave (2020)

Such models as *decision trees* redistribute data at each "leaf", that is easier for the visual comprehension, but may not be easy to explain "why". However, *random forests* using a number of decision trees show high predictiveness (Fig. 2.7).

Since 1990s, extensive research has been aimed at extracting the "rules" generated by trained machine learning algorithms, also to develop methods explaining the behaviour of machine learning processes, how the input correlates with the output, identifying the input–output pairs. Some new more explicable models have been introduced. Among them are the Layerwise relevance propagation, a method for the evaluation of feature impact on a model's decision. Specific transparent models have also been developed, such as Bayesian networks and decision trees. Those methods allow the decision making to be observed and uncover some of the questions about machine learning in general.

[75] Artur D'Avila Garcez, Luis C. Lamb, Neurosymbolic AI: The 3rd Wave, December 2020. p. 2–3 https://arxiv.org/pdf/2012.05876.pdf.

Three Laws of Robotics

A robot may not injure a human being or, through inaction, allow a human being to come to harm.

A robot must obey the orders given it by human beings except where such orders would conflict with the First Law.

A robot must protect its own existence as long as such protection does not conflict with the First or Second Laws.[76]

Isaac Asimov, 1942

On algorithmic transparency

1. A robot must always reveal the basis of its decision
2. A robot must always reveal its actual identity[77]

Marc Rotenberg, EPIC[78] President
James Graves, EPIC Law and Technology Fellow

Though there currently exist different ways to reverse-engineer the decision-making process of a model, and it is even possible to understand with good accuracy which features are the most impactful on the prediction output, this resolves the problem only partially. The current AI research aimed at creating the "explainable AI" (XAI) able to elucidate how the decisions are taken, and thus increasing trust in the AI application. One of the examples is the DARPA Explainable AI (XAI) programme aimed at developing explainable models with high level of performance. It is expected that they will be able to communicate with the users to explain the rationale behind the taken decisions, as well as their behaviour (Fig. 2.8).

Any AI system has its life cycle, that passes through the **development stage** (planning, data collection, designing and creating a model, evaluation, etc.), **validation**, **deployment**, **operation** and **maintenance**, **retirement**.

To get a better insight into the process of creating a trained model, we will analyse the machine learning process step by step. It is essential that each of its learning steps be conducted with consideration of its specificity and compliance with requirements to increase the machine learning performance effectiveness, decrease development costs and address specific vulnerabilities.

For the purpose of this book, we will consider in more detail the development stage, specifically data generation, its pre-processing, training, evaluation, prediction, visualization (see Fig. 2.9).

---

[76] Asimov [14], p. 94.

[77] U.S. Congress, Senate Subcommittee on Space, Science, and Competitiveness, Committee on Commerce, Science, and Transportation, Hearing on the Dawn of Artificial Intelligence, Statement of, 114th Congress, Second session, November 30, 2016, Annex, Letter of EPIC to the Congress, U.S. Government Publishing Office, Washington, 2017, p. 54, https://www.govinfo.gov/content/pkg/CHRG-114shrg24175/pdf/CHRG-114shrg24175.pdf.

[78] EPIC—The Electronic Privacy Information Center.

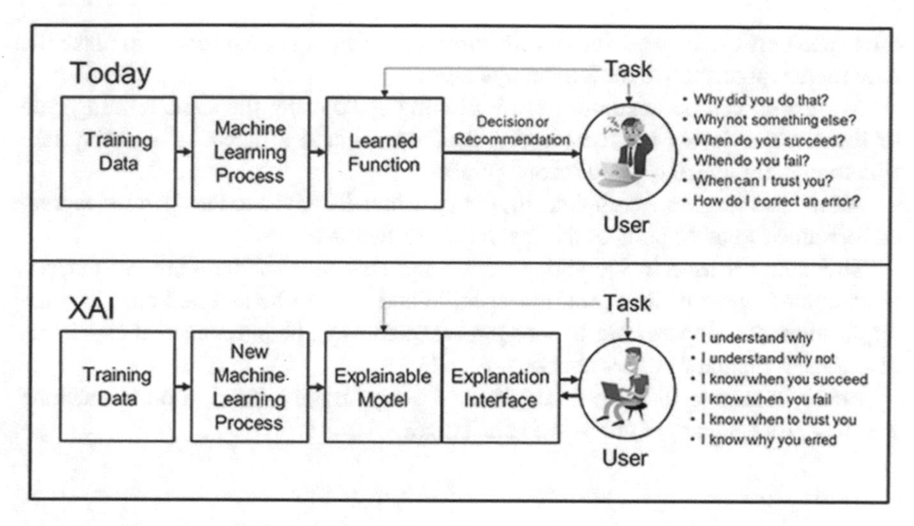

Fig. 2.8 XAI concept. *Source* DARPA web-site[79]

Fig. 2.9 Steps of machine learning

The stages of production, post-deployment use and maintenance of machine learning algorithms will be also covered throughout this book.

Data collection, or generation, being the ground for further patterns development and decisions taking, can be done manually or automatically. For greater accuracy, data collection should be supervised, and parameters set by human operators.

Pre-processing is one of the most important and labour consuming processes consisting in data cleaning, refining, structuring, etc. For supervised and semi-supervised methods labelling is required. Machine learning models require numeric input in a specific range.

Model training, the core learning process, is the most computationally intense, and the requirements vary based on the selection of the model and the quality of the data set.

At the *prediction (or testing) stage* the model is tested using a dedicated "testing" dataset for its capacity to generate predictions. It should be able to guess the output value based on the previous experience.

Predictions are made in a numeric format, which has to be translated into data that can be used for further decision making or statistical analysis. Once processed,

[79] Explainable Artificial Intelligence (XAI), DARPA, https://www.darpa.mil/program/explainable-artificial-intelligence.

those numbers can be used for *visualisation*, e.g., diagram composition, to make the data more representative for human operators.

Evaluation can be done after every training cycle using the same training data or the new previously not used testing data, and in case of needs—further system fine-tuning, optimisation, and reconfiguration.

After those stages are complete, the data is either displayed to the operator, stored, of forwarded to other parts of the system for further use.

The methods used in the above steps vary, and none of them are yet perfect. Furthermore, most of them are task-specific and have to be selected based on the application. Developers have to account, that errors can be introduced at any stage, also though malicious cyber attacks.

Below is a more detailed description of every step, its input into the machine learning process and yet unresolved challenges.

2.3.1 Data Collection

Machine learning starts with the data collection that is fundamental to any machine-learning-based approach. Machine learning models require an adequate training data, to identify and learn the underlying patterns in the data and generalize for predicting the target output of the previously unseen inputs. In an ideal scenario, the *learner* has to extract *knowledge* from the raw data and create a *formula,* or a pattern, that describes how certain features of the data represent some event. For example, how changes in the network traffic flow might indicate a potential attacker trying to infiltrate an industrial network.

Data collection can be manual, automated, or a mix of both. Raw data can be collected from the activities of users, by recording traffic or specific activities on the computer or other systems. For greater accuracy, it is very important to set data parameters, collect the raw data from reliable and correctly filtered sources selected as per the problem to be resolved. The data that misses the set parameters, or have too wide ranges, or have incorrect or corrupted inputs, should be removed. In other words, the raw data should be cleaned.

There is no standard for sizes, dimensionality, or volume of the training and testing datasets. The size of the training datasets in the state-of-the-art methods can be as small as two instances or as large as several million or more instances. However, increasing the size of the training dataset reduces the risk of model over-fitting and improves the model's generalization ability.

Once the data is collected, it has to be converted into **a data set,** that is already the first step of its pre-processing. Thus, creation of data sets is a cross-step process. They can be in a form of a text file, a data base of values in a specific order, or real-time data streams; and usually contain columns of features, that have specific values. Similar to the description of a person using metrics, the network traffic has unique identifiers, like a number of network packets, or the overall volume of a communication session, number of connection breaks and retries, etc. In most of the cases, a data set like that cannot be directly used for training, as it has to be processed and non-numeric values have to be converted into numbers, sometimes even in a specific range between 0 and 1.

Over the past decades the research community has collected, generated and made public a wide variety of datasets for image recognition, sound recognition, medical data, consumer data, etc. Data sets can be downloaded from the research institutions or from the repositories of the machine learning enthusiasts.[80] Cyber security has a dedicated collection of data sets, which contain traffic recordings, malware samples, URLs, emails, or already pre-sorted data, like features of traffic sessions, that can be easily used for experimental intrusion detection systems.

Depending on the set parameters/attributes, the existing datasets may contain:

- Network traffic
- Malware
- Web Applications
- URLs and Domain Names
- Host
- Email
- Fraud
- Honeypots
- Binaries
- Phishing
- Passwords.

To develop security models, specific data should be collected according to best practices to facilitate the development of accurate solutions. Data types depend on the set security service criteria, and can be the following: firewall logs, IDS logs, packet data, network traffic flow metadata, and honeypot data, etc. Currently, this data is collected from the network traffic records, IDS (e.g., Snort), or honeypots.

Finding a suitable data for machine learning in cyber security presents a challenge, especially for some specific types of attacks, also for the most advanced ones. Specific data responding to the set of exclusive criteria, may be generated through attacks simulated for the analysis purposes. Attack-specific data sets may include behaviour of specific malware from the past, for example ZeuS botnet, or phishing emails. Botnet traffic records can be used for detection of very specific attacks. A collection of phishing emails can be dissected to develop anti-phishing systems, since the pre-machine-learning systems can no longer cope with the influx of phishing and spearphishing attacks.

With the growing volume of information transferred between devices, the data collection for cyber security purposes should be reduced to the necessary minimum

[80] Stacy Stanford, "The Best Public Datasets for Machine Learning and Data Science," Medium.com, October 2, 2018, updated 7 August 2020, at https://medium.com/towards-artificial-intelligence/the-50-best-public-datasets-for-machine-learning-d80e9f030279; The 50 Best Free Datasets for Machine Learning, Lionbridge AI, https://lionbridge.ai/datasets/the-50-best-free-datasets-for-machine-learning/;Machine Learning and AI Datasets, Carnegie Mellon University, https://guides.library.cmu.edu/c.php?g=844845&p=6191907.

not to slow down the system performance, e.g. the analysis of every packet. Figuratively speaking, instead of opening every "envelope" and reading the contents, the machine would analyse the "envelope" (i.e., metadata). Metadata contains values that describe traffic flow, sessions, volume of traffic, exchanged between machines. This means, the metadata has to be collected, values calculated, and presented in a format that a machine learning model would understand.

A separate challenge for data scientists is collection of Big Data that is permanently changing with constant in-flows of new data. Due to its high volume, it is impossible to take an accurate snapshot or create a backup in a single given moment. Only separate smaller parts of it can be recorded during a single timeframe, and then assembled into a joint dataset.

The most well-known and researched dataset is KDD Cup 1999,[81] developed by the University of California, which has been further replaced by NSL-KDD. It contains several recorded network attacks including several types of DoS, portsweep, and password bruteforce attempts. Another well researched dataset, developed by the Canadian Institute for Cybersecurity, University of New Brunswick, is CIC-IDS 2017.[82] A less researched dataset CIC-IDS 2018 is also available. University of Victoria has generated and published several specialised datasets, including fake news and biometric datasets.[83]

While collecting the data, it is essential to ensure its high quality and protect it from being corrupted or poisoned. A perfect dataset would have complete knowledge about the subject it is meant to describe. The higher the quality of the collected data is, the more efficient will be the training process and hence, the system performance. The collected data may contain insufficient or even controversial knowledge, it may contain duplicates, biased information, errors, empty spaces, division by zero. This should be addressed to ensure the set parameters are respected. In other words, the data has to be pre-processed.

2.3.2 Pre-processing

Data pre-processing is a mandatory preparatory step in machine learning. It allows to filter, refine and structure the collected raw data. Without pre-processing, a machine learning model simply will not be able to operate. It is worth noting, however, that currently many online datasets are published in a pre-processed format and are distributed as spreadsheet files. However, their verification is required.

The collected data can be in a raw non-structured format, semi-structured, or structured in datasets that may also suit the needs, that is normally a very rare case.

---

[81] KDD Cup 1999 Data, http://kdd.ics.uci.edu/databases/kddcup99/kddcup99.html.

[82] IDS 2017, University of New Brunswick, 2017, accessed on 27 March 2021, https://www.unb.ca/cic/datasets/ids-2017.html.

[83] Datasets, University of Victoria, accessed on 14 June 2021, https://www.uvic.ca/engineering/ece/isot/datasets/.

Its format subsequently defines the amount of effort for its pre-processing. This effort may require the data transformation, normalization, grouping, collation, refinement, application of more specific techniques, e.g. natural language processing.

Not only quality, but also the data quantity, its size, contribute to the performance effectiveness in cyber security. The experiments conducted show that the less the amount of data, the less time is needed for the machine learning. However, it may result in the accuracy reduction in the intrusion detection systems trained with the deep learning techniques.[84] Numerous approaches have been introduced to be able to reduce the size of the datasets through either selection of the most impactful features or by engineering coefficients, that can be used instead of a larger number of features (e.g., Principle Component Analysis). Dimensionality reduction techniques also allow researchers to detect previously unknown correlations between features and derive the value and impact of each feature on a decision-making process. For example, if in a dataset all the attacks originate from one IP address, then the source IP address will be the most impactful feature. This will most likely lead to a bias and all traffic originating from that dataset be treated as malicious. Hence, selected features always have to be verified by developers.

It is not an easy task to determine the level of data sufficiency, its representativeness and quality, and multiple experiments and adjustments may be required. For example, the CIC IDS 2017 dataset contains only a handful of samples of "Infiltration" attacks, while also containing hundreds of thousands of samples of DoS attacks. This dataset would be sufficient to train a model for the detection of DoS attacks, but insufficient for the detection of infiltration attacks.

Data pre-processing is a labour-intensive and time-consuming process, especially in the data labelling for the supervised learning. The existing tools for automatic pre-processing attempt to fix inconsistencies and missing data in the data set, and then convert the data set into a dataframe of numeric values. However, a human supervision is still an essential part of this step. When the pre-processing is ignored, there is a risk of having the unrefined data that will result in a lower accuracy of the output.

Natural language processing

Typical approaches to data pre-processing are narrowed down to the conversion of any data into numeric values between 0 and 1. But natural language is different from any statistical data, as words and sentences have context. To achieve the accurate representation of context, natural language of words, commands, images, lines of code have been converted into various combinations of numbers and values.

[84] Stanislav Abaimov, Towards a Privacy-preserving Deep Learning-based Network Intrusion Detection in Data Distribution Services, Technical report, 12 June 2021, https://arxiv.org/ftp/arxiv/papers/2106/2106.06765.pdf.

Natural language processing has also been successfully applied to the analysis of lines of code and text queries in computer systems. Various types of conversion and masking have been applied, as well as deep learning, support vector machines, and very popular for natural language processing—variations of sequence-to-sequence models.

Pre-processing itself is a complex task that may include several stages. For example, data has to be cleaned, shaped, unified, and eventually encoded.

As traffic may contain personal information and sensitive credentials, the issue of data privacy presents another challenge yet to be resolved. The experiments conducted by one of the authors[85] with the industrial communication protocols, have demonstrated that the use of techniques for the preservation of data privacy reduces the systems effectiveness in attacks detection by a noticeable margin from almost 100% down to 70–80% on average (Abaimov 2021).

With the pre-processed data, machine learning moves to the next training step.

2.3.3 *Training*

Training, also known as learning or fitting, is the core machine learning step, where all essential transformations take place: from the input into the output.

As mentioned previously, the major learning methods are supervised, unsupervised, semi-supervised learning and reinforcement. Semi-supervised learning, as a mixed variation, makes use of the supervised and unsupervised techniques. The difference in those approaches are in a way how the model is guided during training, if the data is labelled or not and if the expected results are provided to it. Reinforcement learning is a technique that allows the model to learn through the guided adaptation in the environment without the knowledge of correct answers or any other additional information.[86] Deep learning technique can be used in any of these methods.

A more detailed description of these methods and techniques will be presented below.

2.3.3.1 Supervised Learning

Machine learning is a thing-labeller, essentially.[87]

[85] Note: by Stanislav Abaimov.

[86] Sutton and Barto [15].

[87] Cassie Kozyrkov, The simplest explanation of machine learning you'll ever read, 15 October 2019, at https://www.linkedin.com/pulse/simplest-explanation-machine-learning-youll-ever-read-cassie-kozyrkov.

Cassie Kozyrkov, Chief Decision Scientist, Google

Definition

Supervised learning is the machine learning task of learning a function that maps an input to an output based on the example input–output pairs.[88] It is assumed that the input data selected for training contains some meaningful patterns that can define the output. It implies that the data sets contain correct answers, that can be presented to the model, and then used to test the accuracy of the system at a later stage of machine learning.

All data in this approach should be initially labelled, either by human operators or other machines. A model infers (selects coefficients) an equation from the labelled training data consisting of a set of training examples. In this type of learning, each example is a pair consisting of an input object (typically a vector) and a desired output value (also called the supervisory signal). The improvements through a series of trainings can enhance the outputs accuracy. Thus, the model can be trained using the same dataset over and over, to improve the accuracy.

As mentioned previously, a high-quality representative dataset is a necessary condition for the learning process. Supervised learning can demonstrate a very high-performance level if trained with the sufficient data that should include the necessary number of labelled examples, in other words be representative. The trained AI system will subsequently be able to provide its own labelling.

Depending on the needs, set goals, costs, available human resources, different type of models are selected and trained.

Classification

Specific algorithms used in supervised learning, may be identified as *Classification* and *Regression.* Some authors add forecasting algorithms and similarity learning as separate approaches; however, they are the applications of classification and regression.

Classification algorithms are applied when as an output the model has to choose from the selected set of values. For example, for the classification of the network traffic flow, the input data is a traffic session or a set of packets, while the output may be the attack type or the traffic classified as non-malicious. In the anti-spam filtering, the model must analyse the contents of the email and its header and decide how to mark the email, "spam" or not.

Regression algorithms are used when the outputs have to be numerical values within a certain range, for example any value from 0 to 1. The model must estimate the correlation as a pattern of input variables. The key difference between regression and classification is that regression focuses on a single "dependent" variable, while other

[88] Stuart and Peter [16].

variables are changing. This makes regression extremely valuable for the analysis of potential events, specifically in forecasting, or making predictions with a degree of probability.

Forecasting and *similarity learning* are specific applications of supervised machine learning, that are sometimes misrepresented as alternative types of it.

Forecasting makes predictions using the knowledge derived from the data. It can be used to predict financial trends, epidemic spread, impact of cyber attacks, timelines, etc.

Similarity learning uses the similarity function to measure the similarity value between objects. It is applied in recommendation systems, and verification methods (e.g., face, voice, gesture). In cyber security, it can be applied to detect how similar traffic samples are with specific attack samples, thus detecting malware behaviour or other type of intrusion that cannot be detected with the signature-based IDS.

Classification and regression can be performed by a variety of models, such as neural networks, Support Vector Machines (SVM), and Hidden Markov Models (HMM). In turn, they can be split into two groups: deterministic and probabilistic.

Deterministic Models

Deterministic models would aim to find a specific range in which the values of features provide the desired answer. Among them are Support Vector Machines (SVM), Liner and Logistic Regressions, etc.

"Support vector machines (SVMs) are particular linear classifiers which are based on the margin maximization principle. They perform structural risk minimization, which improves the complexity of the classifier with the aim of achieving excellent generalization performance. The SVM accomplishes the classification task by constructing, in a higher dimensional space, the hyperplane that optimally separates the data into two categories."[89]

A linear regression model is a method to describe the relationship between two or more variables—a dependent variable, and one or more independent variables.

Logistic regression is an approach to using a logistic function to model a binary dependent variable.

Deterministic models "draw a line" between each category of provided data samples (Fig. 2.10). The number of categories is defined by the number of different labels in the dataset. Deterministic models perform very well in binary classification, but not as good in multi-class classification. As the number of samples grows, the time requirements significantly increase, as deterministic models have limited scalability.

Probabilistic Models

Probabilistic models would yield a probability of two or more output values. For example, in a binary classification the model may predict that the session is 90%

[89] Vapnik [17].

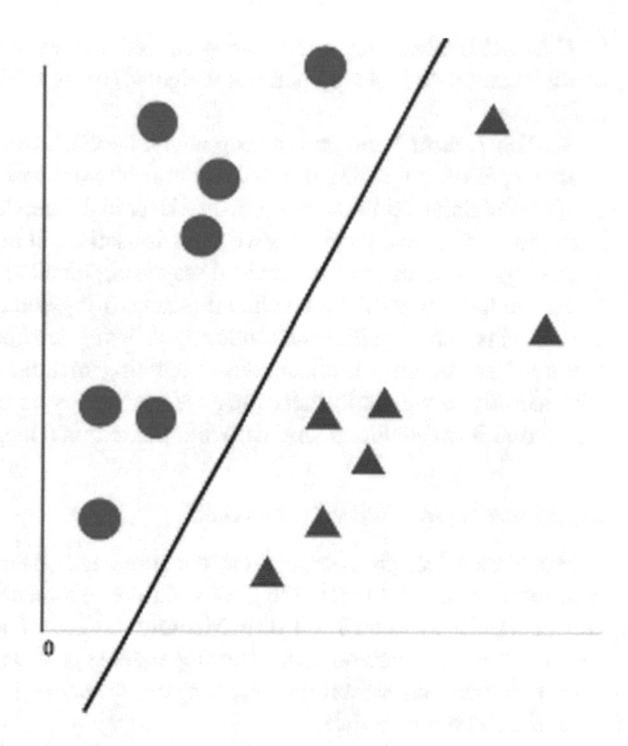

Fig. 2.10 Deterministic classification

normal traffic and 10% malicious. If there are three options, the model may predict 5% normal traffic, 55% DoS attack, 40% port scan.

Among these models there are Naive Bayes, Random Forest, Hidden Markov Models, Decision Tree Learning, K-Nearest Neighbour, etc.

Naive Bayes is a classifier, based on applying Bayes' theorem with strong (naïve) independence assumptions between the features in the data stream.

Decision Tree is a tree-like model of outcomes and their possible consequences, including chance certain events.

"*Random forests* are a combination of tree predictors such that each tree depends on the values of a random vector sampled independently and with the same distribution for all trees in the forest."[90]

Hidden Markov Model is a stochastic model used to model pseudo-randomly changing systems, in which the system being modelled is assumed to be a Markov process—a sequence of possible events in which the probability of each event depends only on the state attained in the previous event.

---

[90] Breiman, L. Random Forests. *Machine Learning* 45, 5–32 (2001). https://doi.org/10.1023/A:1010933404324.

K-Nearest Neighbour methods assign a predicted value to a new observation based on the plurality or mean (sometimes weighted) of its k "Nearest Neighbours" in the training set.[91]

Shallow Neural Network is a deep neural network with only one hidden layer.

In the pre-deep learning times, the neural network was trained one layer at a time (Hinton and Salakhutdinov 2006). Every layer in the network was extracting features from the dataset. Every consecutive layer is used as an input for the next layer. The output layer is trained in a supervised way using labelled dateset. As soon as all the layers are trained, they are combined (stacked) together, to be used as a "model". The weights can be initialised randomly or using unsupervised learning methods. If the dataset is limited, unsupervised learning method could be more beneficial. Alternatively to weight initialisation, the "early" layers of a neural network can be transferred from another neural network, trained on a larger dataset.[92]

Deep Learning in a Supervised Method

In case of deep learning, the supervised learning uses deep neural networks (DNNs), recurrent neural networks (RNNs), convolutional neural networks (CNNs), deep brief networks (DBNs), Long Short Term Memory (LSTM). The latest may be used also in a semi-supervised method. Deep learning models provide more flexibility with their feature engineering capabilities, allowing to extract more knowledge about data, than other classification models.

Fully-connected Feedforward Deep Neural Networks (Multi-layered Perceptron) is a deep neural network in which every neuron is connected to all the neurons in the previous layer. Without any assumptions on the data, this type of deep neural networks is a general-purpose model for classification problems with high computational costs.

In *Convolutional Neural Network* (CNN) each neuron receives its input only from a subset of neurons of the previous layer, to make CNN effective at spatial data analysis. CNN have a lower computation cost than a feedforward network. They are stacked with alternate convolutional (to extract features) and pooling layers (to enhance the feature generalizability).

In *Recurrent Neural Network* (RNN) the neurons send their output values both to following and previous layers. They better perform for sequence data analysis and generation. The *Long Short-Term Memory* (LSTM) model is a variation of RNN, proposed by Hochreiter and Schmidhuber in 1997. It consists of a "forget" gate, an "input" gate, and an "output" gate. Those gates then used to generate the "current" memory state.

---

[91] Fix, Evelyn, and Joseph Lawson Hodges. "Discriminatory analysis. Nonparametric discrimination: Consistency properties." *International Statistical Review/Revue Internationale de Statistique* 57, no. 3 (1989): 238–247.

[92] Alpaydin [13].

Table 2.1 Advantages and challenges of supervised learning

Pros	Cons
• More accurate in task-specific applications • More control over the final outcomes • Less data required • Data can be continuous and discrete • New features can be created	• Requires a labelled dataset with specific categories • Has an issue with overfitting • Time and labour-consuming

Deep Belief Network is a model that consists of several Restricted Boltzmann Machines. Deep Belief Network is typically used for pre-training and feature extraction. An unlabelled dataset can be used optionally, making it an unsupervised or semi-supervised learning method as well.

Among the notorious achievements of the CNN and LSTM are their contribution to enhancing computer vision[93] and speech recognition. There also exist the already established practices for combining different types of layers together. For example, a popular class of convolutional neural networks is ResNet.

Application

Classification models are commonly used for the intrusion detection systems, that have to clearly state if the traffic, email, or file are anomalous or not, and potentially assign it to a specific group.

Regression models are used for binary and multi-class classifications, probabilistic systems, which measure potential probability of the traffic, email, or file to be malicious or not.

However, it is also possible to use classification for predictive purposes, while regression can be used for deterministic problems and predicting continuous values.

Advantages and Disadvantages

Supervised learning has a number of advantages and a few disadvantages (see Table 2.1).

Models can be trained based on the data that has to be labelled and prepared in a specific way. In cases of insufficient data, the model can be trained using the same dataset multiple times, which improves the accuracy, but at a certain point causes overfitting. The equation in the overfitted model becomes too complex and the model accuracy for the new predictions becomes lower than optimal.

---

[93] Krizhevsky, A., Sutskever, I., and Hinton, G. ImageNet classification with deep convolutional neural networks. In Proceedings of NIPS'2012.

Starsky Robotics: autonomous vehicles and supervised learning

In 2015, I got obsessed with the idea of driverless trucks and started Starsky Robotics.

In 2016, we became the first street-legal vehicle to be paid to do real work without a person behind the wheel.

In 2018, we became the first street-legal truck to do a fully unmanned run, albeit on a closed road.

In 2019, our truck became the first fully unmanned truck to drive on a live highway.

And in 2020, we're shutting down.

There are too many problems with the AV [autonomous vehicles] industry to detail here: the professorial pace at which most teams work, the lack of tangible deployment milestones, the open secret that there isn't a robotaxi business model, etc. The biggest, however, is that **supervised machine learning doesn't live up to the hype. It isn't actual artificial intelligence akin to C-3PO, it's a sophisticated pattern-matching tool**.

It's widely understood that the hardest part of building AI is how it deals with situations that happen uncommonly, i.e. edge cases. In fact, the better your model, the harder it is to find robust data sets of novel edge cases. Additionally, the better your model, the more accurate the data you need to improve it. Rather than seeing exponential improvements in the quality of AI performance (à la Moore's Law), we're instead seeing exponential increases in the cost to improve AI systems—**supervised ML seems to follow an S-Curve**.[94]

Stefan Seltz-Axmache

CEO and Co-Founder, *Starsky Robotics*, a driverless truck startup closed in 2020

Future Development

It is expected that a better feature engineering will allow to extract more data from the same dataset, as compared to the older models. This will enable better learning from smaller datasets.

As the data is collected and accumulated, better, larger and already labelled datasets will be published over time. Thus, it will reduce the time on its labelling and pre-processing.

Models can be trained on large datasets using powerful computers, with the subsequent publication of pretrained models in the public domain. They can be used out

[94] Stefan Seltz-Axmacher, The End of Starsky Robotics, 19 March 2020, https://medium.com/starsky-robotics-blog/the-end-of-starsky-robotics-acb8a6a8a5f5.

of the box for the development of a new software, or they can be further trained for a specific task.

Supervised learning provides more control over the final outcome of the model training. The model accuracy can be controlled, and training can be stopped at any moment. Supervised learning will keep improving and create stable methods for the creating of template models, that can be used in the future for new methods of unsupervised and semi-supervised learning.

2.3.3.2 Unsupervised Learning

> True genius resides in the capacity for evaluation of uncertain, hazardous and conflicting information.
>
> *Winston Churchill*

This type of learning is exercised using unlabelled data. Instead of providing desired answers, a model is to decide by itself how many categories (clusters) there are in the dataset, and how they should be identified. As this type of learning does not require predetermined solutions, it might sound like a promising alternative to the supervised learning. However, as we will see later, unsupervised learning is limited in applications and can only be used for clustering or dimensionality reduction.

Definition

Unsupervised learning is a type of machine learning that looks for the previously undetected patterns in a data set with no pre-existing labels, and with a minimum of human supervision. Goodfellow et al. present a definition of unsupervised learning[95] as a process, during which an algorithm learns by itself "to make sense of the data without [the] guide". In contrast to the supervised learning, it self-organizes data, and allows for modelling of probability densities over inputs (see Fig. 2.11). Dense groups of values are categorised separately and are called clusters.

Figure 2.11 shows the example of a random unclustered data, which then gets analysed, grouped, and marked with colour for visual representation.

As in supervised learning, the unsupervised process requires prepared data sets or a data stream to learn from. However, the detected categories are not limited, and different algorithms may detect different number of clusters in the same dataset.

During the training process, the data is processed, selected, structured and categorised based on specific similarities of patterns. The required data sets should be larger than those for the supervised learning, hence the outputs may be less predictable. It may show interesting results when used for the specific purpose of

[95] Goodfellow et al. [18], Deep Learning (Adaptive Computation and Machine Learning series). In Nature (Vol. 521, Issue 7553). https://doi.org/10.1038/nmeth.3707.

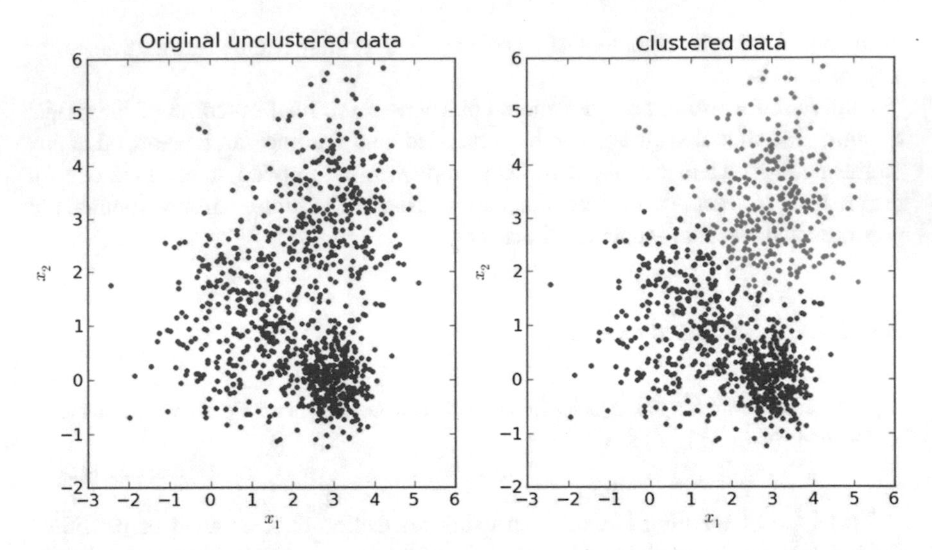

Fig. 2.11 Data before and after clustering

data analysis, finding the unexpected patters specific to the selected data, and in investigating if multiple co-dependent features can be reduced to one.

The unsupervised learning cannot be considered as inferior to the supervised method, as it serves specific purposes where it has relative advantages. For example, a common cyber security application is anomaly detection in network traffic. While the supervised learning can classify the known types of malware or attacks, the unsupervised Learning is able to detect completely new unidentified patterns of behaviour that might indicate unknown types of attack techniques or malware.

Classification

Unsupervised learning groups data through *clustering* and *association.*

In *clustering*, the data is grouped as per its similarity and analysed by relevant characteristics. The approaches for clustering may differ in methods, processing data using various criteria defined by a *similarity value* and *separation.* Those values are then further evaluated by mathematical methods to show the density and the measurable difference between clusters.

Clustering approach represents a flexible solution for the detection of similarities and grouping of objects. It can be used for data analysis or a pre-processing.

Association aims to detect the patterns (known and unknown) between the data features and samples through the sequence identification. Unlike in prediction methods of supervised learning, association outputs more false data and requires a human operator for verification and decision making.

Those two approaches might seem similar, but clustering is focused more on grouping the entire provided dataset into groups, while association is more focused on finding items in the dataset that are similar to a specific query.

The most well-known unsupervised method is *K-means*. It is a very accessible clustering algorithm, present in many programming libraries. "*K*" in "K-means" stands for the number of clusters, while the "means" is the mean of values. The K-means algorithm uses distance as a similarity measure criterion. The shorter the distance between two data objects is, the more likely they are to be placed in the same cluster. It adapts well to linear data. In addition, the K-means requires trial and error as it is sensitive to the parameter K, which is usually set manually before the start.

Restricted Boltzmann Machine (RBM) is a randomised neural network in which units obey the Boltzmann distribution. An RBM is composed of a visible layer and a hidden layer. The units in the same layer are not connected; however, the units in different layers are fully connected. RBMs do not distinguish between the forward and backward directions; thus, the weights in both directions are the same. RBMs are unsupervised learning models trained by the contrastive divergence algorithm, and they are usually applied for feature extraction or denoising. Also used for the unsupervised pretraining of *Deep Brief Network* (DBN).

Deep Belief Networks (DBN) are modelled through a composition of Restricted Boltzmann Machines. DBN can be successfully used for pre-training tasks because they perform feature extraction. They require a training phase, but as the unlabelled datasets can be used, this method is classified as unsupervised learning.

Stacked Autoencoders. They are composed by multiple Autoencoders, a class of neural networks where the number of input and output neurons is the same. Autoencoders excel at pre-training tasks similarly to DBN, and achieve better results on small datasets. An **autoencoder** consists of two models, an encoder and a decoder. In case of deep learning, those can be convolutional and deconvolutional models.

A generative adversarial network (GAN) is a class of machine learning frameworks designed by Ian Goodfellow and his colleagues.[96]

Autoencoders and *Generative adversarial networks* will be discussed in more detail in Sect. 2.3.3.6 *Combining multiple models.*

Application

In the beginning of the 2000s, it was expected that "the main application of unsupervised learning is in the field of density estimation in statistics, such as finding the probability density function."[97] Current applications of unsupervised learning for cyber security include anomaly detection and feature identification. Unsupervised learning may be especially valuable in identifying previously unknown patters, models,

[96] Goodfellow et al. [18], Deep Learning (Adaptive Computation and Machine Learning series). In Nature (Vol. 521, Issue 7553). https://doi.org/10.1038/nmeth.3707.

[97] Jordan and Christopher [19].

Table 2.2 Advantages and challenges supervised learning

Pros	Cons
• Does not require labelling • May identify previously unknown correlations • More autonomous	• May select unintended features for clustering • Requires more data than supervised learning • May introduce errors with continuous learning • Limited scalability

associations in the raw data, in the so-called knowledge discovery, dimensionality reduction.

Data mining

Data mining can be defined as extraction of additional knowledge from the already existing data.

Machine learning and data mining get regularly confused due to the similarities in the assumptions. For the purpose of clarity, it is worth noting that machine learning is evaluated and measured by the ability to reproduce known knowledge, while data mining is evaluated by the ability to detect previously unknown knowledge. However, they may both employ each other's methods.

Advantages and Disadvantages

Unsupervised learning is a constantly developing approach to machine learning. As this approach requires no prior labelled data for learning, it provides a higher degree of flexibility in data selection and its amount, while being effective in discovering hidden patterns and knowledge. It has the benefit of not requiring prior "correct answers," and continuous training. It may also help to identify the best number of defined categories, that can be further used in supervised learning. However, it has disadvantages as well (see Table 2.2). It requires large amount of data, may select unintended features for clustering, has limited scalability, etc.

Future Development

It is a promising area of research and application and with technology becoming more accessible, the interest in unsupervised learning grows. As new data is being collected and processed, new models are being pre-trained.

A potential direction for unsupervised learning is a broader application of its ability to decide the number of clusters independently. Another direction will be revealed, once unsupervised learning is advanced enough to extract more knowledge without prior indication of where to look. This will bring machine learning one step closer to autonomy.

2.3.3.3 Semi-supervised Learning

Definition

As the name suggests, semi-supervised learning is an approach that uses techniques and combines characteristics of both supervised and unsupervised learning. It is assumed that the combination of the unlabelled and labelled data in various proportions may produce better results.

Classification

As semi-supervised learning is a combination of methods from supervised and unsupervised learning, it is not always possible to classify and attribute them in a simple way. Nevertheless, semi-supervised learning can be divided into the following training categories based on:

- Yarovsky algorithm (self-training)
- low density separation methods
- graph-based algorithms.

Semi-supervised learning can be applied to problems of classification and clustering. One of the examples illustrating the learning process is the use of the *Deep Belief Network*s (DBN), mentioned earlier. In this particular case, DBNs are semi-supervised, as there are two distinct stages in their training: unsupervised pretraining and supervised fine-tuning. Among their application are feature extraction and classification in cyber attack detection.

Applications with Examples

Semi-supervised learning showed itself as most effective in the Natural language processing, e.g. in machine translation. In this capacity, it can be used for vulnerable code analysis, query analysis, and event notification analysis.

Even using a small volume of labelled data, the semi-supervised learning can be used to pretrain models for intrusion detection, that could be further deployed in any network to learn its behaviour.

Table 2.3 Advantages and disadvantages of semi-supervised learning

Pros	Cons
• Allows to use less labelled data for training • Combines problem solving • Easy to implement • Fast • Considered to be a stable algorithm	• Amplifies noise in labelled data • Requires a balanced labelled dataset • Comparatively low accuracy

Advantages and Disadvantages

Semi-supervised approach, while being slightly more complex in implementation than supervised and unsupervised ones, demonstrates the benefits of both (Table 2.3).

Future Development

Semi-supervised learning has a potential to be one of the valuable approaches to commercially available machine-learning-based systems. A big model can be pretrained using supervised learning and a labelled dataset, and deployed in commercial networks for unsupervised learning about the traffic behaviour. This approach can created "personalised" anomaly detection systems, that can be deployed as a service.

2.3.3.4 Reinforcement Learning

An alternative to semi-supervised learning is the reinforcement learning, which allows the model to take decisions, e.g. to reach from point A to point B autonomously, with its own trial and error and while "receiving rewards/reinforcement signals" for successful decisions/actions, and being penalised for errors.

Popularly applied for video games, reinforcement learning has high potential in robotics and generally in autonomy of systems.

Reward and punishment education

We normally associate punishments and rewards with the teaching process. Some simple child-machines can be constructed or programmed on this sort of principle. The machine has to be so constructed that events which shortly preceded the occurrence of a punishment-signal are unlikely to be repeated, whereas a reward-signal increased the probability of repetition of the events which led up to it.

... The use of punishments and rewards can at best be a part of the teaching process. ... It is necessary ... to have some other 'unemotional' channels of communication. If these are available it is possible to teach a machine by punishments and rewards to obey orders given in some language, e.g. a symbolic language. These orders are to be transmitted through the 'unemotional' channels. The use of this language will diminish greatly the number of punishments and rewards required.[98]

Alan Turing, computer and AI pioneer, 1950

Definition and Characteristics

Reinforcement learning studies the methods and behaviours of software agents, that take actions in an environment in order to maximise the cumulative numeric reward for an accomplished task. "Reinforcement learning is the learning of a mapping from situations to actions so as to maximise a measurable reward or reinforcement signal."[99] Reinforcement model explores the environment for possible solutions or actions that can be taken in an attempt to calculate the most optimal sequence of steps that will yield the highest final reward.

Reinforcement learning is designed around the principle of "rewarding" the model for correct predictions using an unlabelled dataset. In reality, reward or reinforcement signal is a value that is added to a counting variable, while penalty is a value that is subtracted from the same counter.

A model is tasked to choose a path in the real or simulated environment through finding solutions, or building on the lessons learnt from the past experience. Specific parameters, values and goals, are provided for this learning process that is to a certain extent autonomous. Data does not need to be labelled and is collected through environment exploration and perceiving the rightfulness of the action through accumulating "rewards". The only supervision provided is the signal of failure to the model, so it restarts the environmental exploration from the beginning.

The simplified reinforcement learning process is presented in Fig. 2.12.

The simulated environment is typically represented as a set of different states with assigned values and rewards. One specific method frequently used in reinforcement learning to represent this kind of environment is known as Markov decision process (see Fig. 2.13).

[98] A. M. Turing, I. Computing machinery and intelligence. *Mind*, Volume LIX, Issue 236, October 1950, p. 457, https://doi.org/10.1093/mind/LIX.236.433.

[99] Sutton and Barto [15].

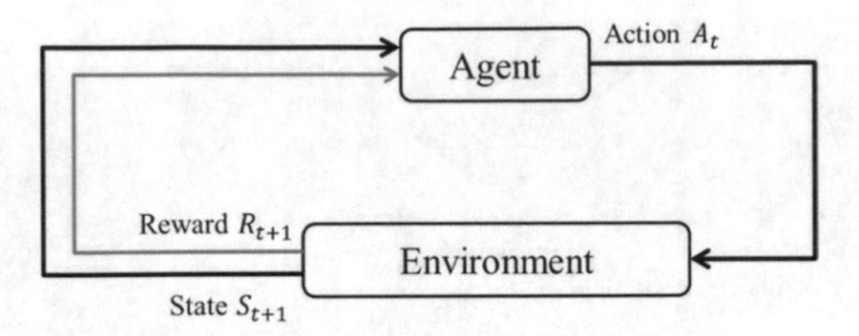

Fig. 2.12 Simplified reinforcement learning diagram. *Source* G. Allen, 2020

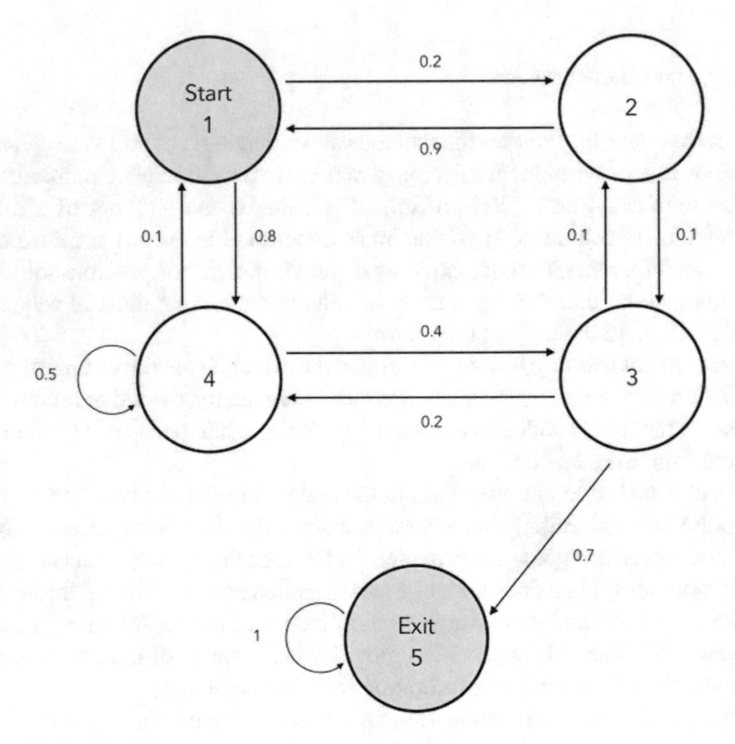

Fig. 2.13 Example of a Markov decision process

Classification

Among the most widely used reinforcement learning methods, there are the following:

- Dynamic programming
- Monte-Carlo methods
- Heuristic method.

Dynamic programming is a method which splits the problem into subproblems and implies that the optimal solution to the final problem is the sum of optimal solutions to subproblems.

Monte-Carlo method is a wide class of algorithms of random sampling that aim to obtain final approximate results of values that would require more complex computations in the setting where accurate results are not essential.

Heuristic method is a method that ranks alternatives using various search algorithms based on available information to decide which action to take at each step.

There have been developed a few specific algorithms for reinforcement learning.

Q-Learning is the most well-known algorithm for reinforcement learning, being also one of the oldest and widely used. Q-learning was proposed by Watkins and Dayan [21] and it is a variation of an older concept—Markov Decision Process (see Fig. 2.13). The model repeatedly attempts to traverse every possible combination of steps and learns from the simulated or real environment to gain gradually higher total values (scores). As soon as the model manages to complete the task by the most optimal means, measures as the highest score, the model is considered to be trained and a so-called Q-matrix is obtained. Q is a utility function, that is used to calculate the sum of rewards and penalties.

Deep Q-Network is one of the deep learning algorithms and the first deep reinforcement learning method, proposed by DeepMind (Google). It applies Deep Q-Network to Atari games. This approach differs from supervised and unsupervised learning, and uses direct video stream as an input.

The Asynchronous Advantage Actor-Critic Algorithm (A3C) is a deep enhancement learning algorithm proposed by DeepMind in 2016 [22]. A3C completely utilizes the Actor-Critic framework and introduces the idea of asynchronous training, which can improve the performance and speeds up the whole training process. If the action is considered to be bad, the possibility for this action will be reduced. Through iterative training, A3C constantly adjusts the neural network to find the best action selected policy.

Trust Region Policy Optimization (TRPO) is proposed by Schulman in [23], it is a kind of random strategy search method in strategy search method. TRPO can solves the problem of step selection of gradient update, and gives a monotonous strategy improvement method. For each training iterative, whole-trajectory rollouts of a stochastic policy are used to calculate the update to the policy parameters θ, while controlling the change in policy as measured by the KL divergence between the old and the new policies.

Another algorithm, known as **UNREAL** has also been proposed by DeepMind in 2016 for the depth-enhancement learning. Based on the A3C algorithm, UNREAL outperformed a human player at Atari by 8.8 times. There are two types of auxiliary tasks in the UNREAL methods: the control task for character control and the back-prediction task for feedback. UNREAL algorithm uses continuous multi-frame image input from the video stream, in order to predict the next-step feedback value, and uses history information to additionally increase the value iteration task.

Table 2.4 Advantages and disadvantages of Reinforcement learning

Pros	Cons
• Potential to autonomy • Lower costs in data generation • High speed in decision making in the real environment • Suitable for problems where future actions are based on the outcomes of preceding actions	• Errors in real-life environment may occur endangering human lives • Data collected in the learning process may be corrupted and lead to errors • High vulnerability to cyber attacks • Hard to apply for cyber security

Application

Reinforcement learning is applied and researched in many areas, e.g. economics, information theory, multi-agent systems, game theory, statistics, etc. It is used in autonomous and self-driving vehicles, for resource management, in robotics, in playing games, e.g., Go, chess, poker, etc. Its use is advantageous and cost effective in the areas where there are no definite models for branching sequences of actions as it generates data while in action.

In the areas of game playing, Reinforcement learning demonstrates feasible advantages over human reaction. For example, in the system AlphaGo, the model was trained for the board game Go. During the three days of the training stage, AlphaGo played around five million games. The system learnt to play Go at a level exceeding world champions. In real life it takes approximately one hour to play the game, but AlphaGo, in average, spent 19 s for playing and completing one game.

Unlike in board and video games, intrusion detection does not always have a sequence of actions that need to be taken. Only in 2020 reinforcement learning has been formally surveyed in the context of network intrusion detection.[100]

Current Advantages and Disadvantages

The area of application is limited to problems that have branching sequences (trees) of solutions. Table 2.4 summarizes the advantages and disadvantages of this type of learning.

The vulnerabilities of reinforcement learning to cyber attacks are recognized and being widely researched. Even small adversarial changes result in a significant drop in the model performance regardless of the learned task or a training algorithm. For example, models can misclassify images as a result of "Adversarial Attacks" that change the input images almost unnoticeable to the human eye. Furthermore, malicious samples can be injected into one dataset and infect another one, if both models were trained using the same dataset, causing a so called cross-dataset contamination.

[100] Lopez-Martin, M., Carro, B., & Sanchez-Esguevillas, A. (2020). Application of deep reinforcement learning to intrusion detection for supervised problems. Expert Systems with Applications, 141, 112,963. https://doi.org/10.1016/j.eswa.2019.112963.

Future Development

It is expected that reinforcement learning will be further used to advance autonomy in systems, e.g. in self-driving vehicles, alongside semi-supervised learning. Deep reinforcement learning has already been applied to the long-distance route planning and optimal velocities in self-driving vehicles and connected autonomous vehicles. However, it is worth highlighting that at the current stage of the technological development, connected autonomous vehicles are highly vulnerable to data spoofing attacks.

In industrial control system, reinforcement learning is used to address voltage fluctuations. In addition, in robotics reinforcement learning can be used for visual navigation. For mechatronics research, a soft artificial muscle-driven robot cuttlefish uses reinforcement learning for control and pathfinding.[101]

Further research will by all means continue in the original area of the reinforcement learning application—the learning to play games and for video games.

Its use in the open environment remains a challenge, with it being too complex even for the modern generation of sensors. Also, vulnerabilities to cyber attacks strongly undermine the reliability of the systems trained with the reinforcement learning, especially in the critical systems. With these weaknesses successfully addressed and technological advancements—new generation of sensors, enhanced algorithms—the reinforcement learning will undoubtfully become one of the widely used approaches in future.

2.3.3.5 Selecting Types of Training

The previous subchapters have reviewed in detail the specificity of machine learning algorithms and types of learning, their strong and weak sides, and potential challenges. The knowledge of these characteristics can be very helpful in selecting suitable models for developing effective AI systems.

While selecting the type of training it is necessary to keep in mind that the supervised learning is very labour intensive and requires large labelled datasets. The reinforcement learning will be based on multiple interactions that may not be efficient in critical situations. Though unsupervised learning is the least intense in computational requirements, it may have its own vision of the data procession not necessarily corresponding to the expectations.

Table 2.5 provides a comparative summary of the major characteristics of the training methods.

Machine learning is successfully applied in cyber security, and an extensive research is ongoing in its application in intrusion detection systems (IDS). Most used methods for intrusion detection and intrusion classification are Support Vector Machines and the variations of Deep Neural Networks used both in supervised and

[101] Yang, T., Xiao, Y., Zhang, Z. et al. A soft artificial muscle driven robot with reinforcement learning. Sci Rep 8, 14,518 (2018). https://doi.org/10.1038/s41598-018-32757-9.

Table 2.5 Comparative characteristics of major machine learning approaches

Criteria	Approaches			
	Supervised	Unsupervised	Semi-supervised	Reinforcement
Data	Labelled	Unlabelled	Both labelled and unlabelled	Autonomous AI agents that gather their own data and improve based on their trial and error interaction with the environment
Training time	Time-consuming for data labelling	Does not request any time for labelling, but may need more experiments	Depends on the amount of labelled and unlabelled data	Training-time is task-dependent, as the system will require to continuously re-launch and re-attempt simulated instances
Cost	High/Medium[102]	High	Medium	High
Human supervision	Required	Not required	Both	Not required
Performance	High predictability and high performance	Lower predictability and performance, but more effective in anomaly detection	Mixed	High performance for specific applications
Application	for binary and multi-class classifications, probabilistic systems (to measure potential probability of traffic, email, or file to be malicious or not), intrusion detection	For anomaly detection; dimensionality reduction—less traffic; identifying previously unknown patters, models, associations in the raw data, knowledge discovery, intrusion detection	in classification and clustering, Natural language processing, machine translation, fraud detection, intrusion detection systems	in autonomous and self-driving vehicles, for resource management, in robotics, playing games, statistics, economics, etc.

[102] The cost will also depend on the data labelling. If the labelled data is used, it can be less expensive.

unsupervised learning. Reinforcement learning for the network intrusion detection is a new approach, that has not been fully researched yet.

The below **checklist** of a preliminary assessment can help in identifying a specific type of machine learning responding to the needs and requirements.

No.	Questions	Answers	Recommendations
1	Is the dataset collected?	Yes	Pre-process
		No	Download online or collect own data
		No, and doesn't have to be	Unsupervised, Reinforcement
2	Is the dataset large or small?	Large	Unsupervised
		Small	Supervised
3	Is the quality high or low	High	Supervised
		Low	Pre-processing to be reinforced
4	Is the data available or to be collected through simulation?	Available	Standard pre-processing
		To be simulated	Generation with GAN
5	Is there a need to improve the data?	Yes	Data augmentation techniques, GAN
6	Is the dataset labelled?	Yes	Supervised, Semi-supervised
		No	Unsupervised
7	Is the data representative?	Yes	Remove duplicates
8	Will the post-deployment need additional learning?	Yes	Unsupervised, Semi-supervised
		No	Supervised

Development of the machine-learning-based systems is an intricate task and the decision about the machine learning methods have to be made early on. The provided checklist can shed some light and hint on general ideas of what should be done. In each case, the answers to the checklist questions above will guide the developers and project managers in the selection of learning types for a specific subsystem.

2.3.3.6 Combining Multiple Models

The machine learning algorithms are implemented through the model functions and stored variables. During the training models are aimed to reach a peak level of accuracy, and they do not improve further.

An idea of using several models to outperform a single-model system has been around for almost three decades. With the growing availability of computational power, deploying multiple models at once is no longer impossible. It is worth noting, however, that multiple models can not only reinforce, but also compete with each other.

In *ensembles* multiple models work together to improve the quality. But not every multi-model method can be called an ensemble. Thus, in *Autoencoders* and *Generative adversarial networks* there are two models that do the opposite tasks or work against each other.

Autoencoders

An **autoencoder** consists of two models that do the exact opposite tasks, an encoder and a decoder. The encoder attempts to output features from the input dataset, while the decoder aims to reconstruct the original data from the outputs of the encoder. During the training stage, if the decoder reconstructs the data from the available synthetic features, it indicates that those synthetic features accurately represent the original data, only in a smaller scale. When the input features are independent and not correlated to each other, autoencoder will not be able to properly reconstruct the original data. No supervision is required through the entire process.

Probably the most well-known autoencoder application is a denoising autoencoder. Denoising (or noise reduction) is the process of removing noise from a signal. This can be applied to image, video, audio or even text. In cyber security autoencoders help to identify most valuable features for detection of cyber attacks, as they would provide a higher number of discriminative features as compared to other feature engineering methods.

Generative Adversarial Networks

A generative adversarial network (GAN)[103] is a class of machine learning frameworks designed by Ian Goodfellow and his colleagues. In the contest between two neural networks there is a zero-sum challenge, where one model improves a value by the loss of value in another model. GANs generate adversarial examples by using random noise as a source input, and transform it into new samples for the dataset, that are similar to those in the dataset. Exactly same approach is used to generate new images, photos, audio, and video.

Generative approach allows to create synthetic samples for datasets, augmenting and expanding them. The challenge in many tasks is to generate the samples that are functional. For example, generated faces can look almost human, while in the attempt to generate synthetic network traffic or code injection samples, one symbol difference between generated malicious samples mean a difference between a successful exploitation or a crash of the system.

[103] Goodfellow, I., Pouget-Abadie, J., Mirza, M., Xu, B., Warde-Farley, D., Ozair, S., Courville, A., and Bengio, Y. Generative adversarial nets. In Advances in Neural Information Processing Systems, 2014, 2672–26,804.

Adversarial samples are created by introducing small changes to the already existing patterns. The refinement technique uses two models: generator and discriminator. One creates new samples and another one rates them. The main concept at the core of the method is the "indirect" training through the dynamically updated discriminator. The discriminator generates feedback, that refines the randomly generated data into synthetic samples that are indistinguishable from the real data. Unlike traditional models that aim for maximum accuracy, the generator aims to fool the discriminator. The competition between two models make the method adversarial, even though the final goal is to create new "real" data. As long as the process repeats in a continuous cycle, it will keep producing new synthetic data.

GAN was originally proposed as a subcategory of unsupervised learning, but it has also proven itself as applicable in other types of learning.

GANs can be applied to create malware, that can be indistinguishable from non-malicious data or executable files. Alternatively, GANs can be applied for anomaly detection using only non-malicious activity records or normal network traffic. With this method it is possible to evaluate whether the testing sample is a potential cyber-attack or not. The challenge is to assign a numeric value to the fluctuations from the "normal" behaviour or patterns.

Ensembles

Ensembling has already proven to outperform single models, however, it comes with a cost of additional computational resources at the deployment stage.

There are several approaches to how better organise the relations between multiple models, as they have to train with the same dataset. The most widely used ensembles are *bagging, boosting*, and *stacking*. *Bagging* is a technique of training multiple classifiers in parallel and then combining them using any of the averaging methods. *Boosting* is a technique for ensembling sequentially trained classifiers. *Stacking* is a method of training multiple classifiers in parallel and then combining them using another model specifically for making the final decision.

Those three types of ensembles differ in their adjudication method. Adjudication is the step taken after multiple learners provided their predictions. Those predictions have to be then processed, like in voting, to take a single decision, if it is an attack, or a non-malicious operation that has to be allowed. For example, *bagging* uses different averaging techniques for adjudication, while *stacking* uses an additional "output" model for final decision-making.

In an ensemble of "expert" models, each separate model (e.g., neural network) is trained in detecting a different type of cyber attacks, and then their combined efforts increase the overall accuracy of intrusion detection.

The development of machine learning methods using ensembling, can be automated by utilising programming libraries (e.g., SciKit Learn). To ensure full control over the learners, developers may use their own approaches to building ensembles. In automated methods more often than not ensemble functions are limited to a specific type of classifier. For example, in SciKit Learn it is not possible to use custom made deep neural networks for Gradient Boosting with a custom number of learners, only Random Trees can be used.

A single neural network is normally less efficient than an ensemble. For example, an individual Artificial Neural Network, that is typically trained to the highest possible performance, very rarely reaches 100% accuracy. To achieve the highest possible accuracy with a single neural network, substantial computational resources and advanced detailed configurations are required at the training stage. And even then, the results may not be sufficient for an acceptable performance of the designed system. Furthermore, a high-accuracy neural networks can be overfitted and perform poorly with unknown data.

Concerns are equally voiced around the behaviour of ensemble learning with deep neural networks, as those are not yet well studied and understood. The challenge is that the researchers focus mainly on the design of a neural network structure without improving the approaches to ensembling. Unweighted averaging technique, which is most commonly used for adjudication of models' outputs, is not data-adaptive and thus vulnerable to poor selection of base models. Adjudication based on averaging performs well for networks with similar structure and comparable performance, but it is sensitive to the presence of excessively biased (overfitted or underfitted) weak learners.

Below is a more detailed description of the several most well-known ensembles.

Bagging

Bagging (Bootstrap AGGrigating) was first proposed by Breiman,[104] based on the bootstrap sampling method.[105] It consists in the generation of several different smaller datasets drawn at random from the original training dataset, with their subsequent use for training different neural networks. Some samples of original training set may be repeated in the resulting training set while others may be left out. Each model runs independently and the outputs of those models are then adjudicated together using various methods, providing a single final output value or sequence of values.

Bagging leads to "improvements for unstable procedures", which include, for example, artificial neural networks, classification and regression trees, and the subset selection in linear regression. Bagging was shown to improve preimage learning. On the other hand, it can mildly degrade the performance of stable methods such as K-nearest neighbours.

Table 2.6 demonstrates advantages and disadvantages of Bagging.

---

[104] Breiman, L. (1996). Bagging Predictors. Machine Learning, 24(2), 123–140. https://doi.org/10.1023/A:1018054314350.

[105] Efron and Tibshirani [20].

Table 2.6 Advantages and disadvantages of bagging

Advantages	Disadvantages
• Many weak learners aggregated typically outperform a single learner over the entire set, and has less overfit • Removes variance in high-variance low-bias data sets • Can be performed in parallel, as each separate bootstrap can be processed on its own before combination	• Risk of carrying a high bias into its aggregate if dealing with a data set with high bias • Loss of interpretability of a model • Can be computationally expensive

Boosting

Boosting is a process of training several "weak" learners. Each learner, after making predictions, dictates what features the next learner will focus on. More weight is given to samples that were misclassified by the earlier rounds.

Boosting was originally proposed by Schapire[106] and improved by Freund et al.[107,108] It generates a sequence of models whose training sets are determined by the performance of former ones. Training instances that are wrongly predicted by former networks will play more important roles in the training of later networks.

One of its variations that is commonly applied for intrusion detection is AdaBoosting. In Adaboost the samples, incorrectly predicted by the first model, are stored. Then a weight is added to misclassified predictions and the next model makes another cycle of predictions. This way a formed ensemble makes better predictions by using all previous models.

In the Gradient boost method, instead of adding any weights, the samples that were predicted incorrectly are considered as a new training set and the new model. Then predictions are done using new model and the next set of wrongly predicted samples is again used to train a third model. An ensemble then makes better predictions than a single classifier.

Stacking

Stacking uses multiple models and adjudicates the outputs via a meta-model. The models are trained with a full training dataset. The decision-making model is trained on the outputs of the base models as features, to provide the final prediction or classification.

[106] Schapire, R. E. (1990). The strength of weak learnability. Machine Learning, 5(2), 197–227. https://doi.org/10.1007/BF00116037.

[107] Freund, Y. (1995). Boosting a Weak Learning Algorithm by Majority. Information and Computation, 121(2), 256–285. https://doi.org/10.1006/INCO.1995.1136.

[108] Freund, Yoav, & Schapire, R. E. (1997). A Decision-Theoretic Generalization of On-Line Learning and an Application to Boosting. Journal of Computer and System Sciences, 55(1), 119–139. https://doi.org/10.1006/JCSS.1997.1504.

Wolpert[109] was the first to introduce a term "stacked generalisation", which later became known simply as "stacking". Later, Breiman[110] demonstrated that stacking can be used to improve the predictive accuracy in for regression problems, and showed that imposing certain constraints on the adjudication model improved predictive performance.

Stacking constructs two areas of ensemble preparing data and ensemble combination. It selects training data for ensemble units by a cross validation technique, for exploring second level generalisers, it combines the results of the first level models. The data is supplied to the first-level models by partitioning the original datasets, which is divided two or more subsets. Every model is trained by one part of the dataset, and the rest of the part is used to generate the outputs of the learners (to be used as the second space generalisers input). Then second level generalisers are trained with the original ensembles outputs that are treated as the correct guess. In fact, stacked generalisation works by combined classifiers with weights according to the individual classifier performance, to find a best combination of ensemble outputs.

The next steps in the machine learning process are predictions and the evaluation of the final model, before it can be deployed.

2.3.4 Prediction

Prediction is the most valuable stage, as it is the one stage machine learning is known for. It is the reason all the previous steps were needed. When the model is trained to the desired accuracy, it is applied to analyse new input samples that were not in the training dataset. In other words, once a model is trained, it can predict new values, forming a new testing dataset.

As an example ...

A full dataset contains 1000 samples. Out of those, 800 are randomly selected to be used for training and the remaining 200—for testing. Once the model is trained with 800 samples, the trained model (as a programming variable or a file) is used to take 200 testing samples, and makes a prediction for each one of them separately. The output of the model will be 200 different predictions. As the desired outputs are known in advance, they are compared with the predicted 200 outputs, and statistics is calculated for further evaluation.

---

[109] Wolpert, D. H. (1992). Stacked generalization. Neural Networks, 5(2), 241–259. https://doi.org/10.1016/S0893-6080(05)80023-1.

[110] Breiman, L. (1996). Stacked regressions. Machine Learning, 24(1), 49–64. https://doi.org/10.1007/BF00117832.

As mentioned earlier, the output of one or more models can be used as an input to the next model, creating a more complex system, described in Sect. 2.3.3.6.

Depending on the models used, the numeric outputs will be different. They can be a number between 0 and 1, or a set of number combinations that identify a specific classification. Those values have to be interpreted and converted into a format that operators can understand. Values have to be decoded and presented in a readable format. Single values have to be scaled up and decoded as probabilities or measurements. Series of classification values have to be re-assigned to the matching categories as during the input.

Interpretation of those numbers into readable and understandable format is known as post-processing.

2.3.5 Evaluation and Metrics

The testing dataset is used to analyse the performance and predictive capabilities of the model, and to estimate the quality and usability of the model. It has to be evaluated in a numeric equivalent using statistical methods and metrics. For example, Natural Language Processing can be evaluated automatically only if the design is based on binary classification or regression. Otherwise, specific evaluation methods are required. There are several approaches to evaluating Binary and Non-binary Classification or regression problems, but all of them can be evaluated using a few common mathematical methods.

Guessing entropy (GE) represents the average number of guesses needed to reveal the target information. In some studied papers, the used machine learning models are evaluated based on the number of sidechannel attack traces required to achieve a certain fixed guessing entropy.

Success Rate (SR), Recognition Rate, Prediction Rate, Accuracy or Prediction Accuracy represent the ratio of the data points that are correctly predicted to the total size of the data under test.

True Positive Rate (TPR) or Sensitivity or Recall is used in binary classification problems where TPR represents the ratio of the correctly predicted data as positive (TP) to the total number of genuine positive data (GP).

False Positive Rate (FPR) is used in binary classification problems where FPR represents the ratio of the data incorrectly classified as positive (FP) to the total number of genuine negative data (GN).

True Negative Rate (TNR) or Specificity is used in binary classification problems where TNR represents the ratio of the data correctly classified as negative (TN) to the total number of genuine negative data (GN).

Precision is the ratio of the correctly predicted data as positive (TP) to the total predicted data as positive (TP + FP).

ROC curve is a plot of TPR against FPR obtained at different threshold values applied to the classifier scores. ROC curves are a very useful visual indicator of the quality of a model. The area under the curve (AUC), is an aggregated measure of performance of a binary classifier on all possible threshold values (and therefore it is threshold invariant). AUC calculates the area under the ROC curve.

F-measure or *F-score* is the harmonic mean of precision and recall. Precision is also known as positive predictive value, and recall is also known as sensitivity in diagnostic binary classification.

It is recognised that F1-score represents the quality of the model rather than accuracy value. Meanwhile, it is argued that *Matthews correlation coefficient* (MCC) can be more relevant to the metrics, than F1 score.[111] MCC is better suited for classification than for regression evaluation.

Mean Absolute Error is the average of the difference between the Original Values and the Predicted Values. It gives us the measure of how far the predictions were from the actual output. However, they do not give us any idea of the direction of the error, i.e. whether we are under predicting the data or over predicting the data.

Mean Squared Error (MSE) is quite similar to the Mean Absolute Error, the only variance being that MSE takes the average of the square of the difference between the original values and the predicted values. The advantage of MSE being that it is easier to compute the gradient, whereas Mean Absolute Error requires complicated linear programming tools to compute the gradient. As the square of the error is taken, the effect of larger errors become more pronounced then smaller error, hence the model can now focus more on the larger errors.

Keeping in mind that machine learning aims for generalisation of knowledge provided, data itself may cause bias in the algorithm, values, listed above, provide the performance evaluation of a specific model for a specific dataset, but not for the entire application.

After a successful evaluation, the model is either approved for deployment, or a new model has to be trained with a different set of configuration parameters. Once the model is approved for deployment, it can be integrated into the already existing software, or to an entirely new independent program or service.

2.3.6 Fine-Tuning

Fine-tuning is a way to improve the performance or accuracy of the model through the changes in parameters of the training process or the model itself. The incorrectly configured models perform poorly or fail to extract knowledge from data at all. Each type of machine learning model has its unique adjustable parameters, but they also share some common characteristics, such as a number of training cycles or the learning rate.

Fine-tuning speeds up training and improves the performance through eliminating potential errors that may stem from for example incomplete or imbalanced datasets.

[111] Chicco, D., & Jurman, G. (2020). The advantages of the Matthews correlation coefficient (MCC) over F1 score and accuracy in binary classification evaluation. BMC Genomics, 21(1). https://doi.org/10.1186/s12864-019-6413-7.

Training the same model multiple times using the same dataset overcomes the limitations of smaller datasets and improve the accuracy of the model. However, high learning rate may cause overfitting and errors in the learning process, while low learning rate will slow down the training. Thus, *learning rate* can be adjusted to increase or slow down the improvement of the model after each iteration through the dataset.

AutoML and AutoKeras are the examples of Python libraries, that provide automation for fine-tuning.

Machine learning is evolving in parallel with the technologies' development. With the introduction of quantum computers, a new phenomenon is coming on the scene—quantum machine learning.

2.4 Quantum Machine Learning

Conventional binary computers have almost a century of history of evolution, and their performance capacity has almost exhausted itself. In search for innovations, scientists explore other alternatives.

The idea to build a computer based on the principles of quantum mechanics came onto the scene in the 1980s as an outcome of the search for a more powerful way and need to resolve emerging challenges impossible to resolve by the systems operationalised through encoding data into binary "bits"—as ones and zeroes (see Feynman [24]; Deutsch 1985[112]). In the early 1980s, Richard Feynman envisaged that quantum computers and quantum mechanics principles could provide a way to solve problems that are exponentially harder for the contemporary computers, which are based essentially on classical physics concepts and hardwares.

Among the reasons for resorting to the quantum mechanics was the fact that it quantum mechanics is a non-deterministic theory, and that a physical microscopic system of particles and atoms is based on a mathematic principles that involve infinite possible states of that system, which could be described by a sort of matrix or vector having infinite enteries belonging to a Hilbert space. Furthermore the basic principles of quantum mechanics are non-intuitive, but, pragmatically saying, work in the microcosmos and can be experimentally proven. In a seminal paper titled "Simulating Physics with Computers",[113] the Nobel Laureate Richard Feynman proposed in 1981 that in order to simulate quantum physics one needs a radically different

[112] Deutsch David, Quantum theory, the Church–Turing principle and the universal quantum computer, *Proc. R. Soc. Lond.* A40097–117, 1985, https://doi.org/10.1098/rspa.1985.0070.

[113] Feynman, R.P. Simulating physics with computers. *Int J Theor Phys* 21, 467–488 (1982). https://doi.org/10.1007/BF02650179.

concept of bit, as well as a universal computer known as Universal Turing Machine on which all the existing computation is based.

Quantum computers are the next generation technology that is using the achievements of the quantum mechanics. They are characterised by enhanced computational capabilities and possess large storage capacities.

2.4.1 Quantum Computers

This new generation of computers is able to "use the properties of quantum physics to store data and perform computations."[114] These are not only the machines that have increased computational power, using quantum phenomenon, such as superposition and entanglement[115] to perform computations, they are working in a mode completely different from conventional computers. Those quantum properties result in new ways of data representation, higher speed, quantum cryptography, quantum machine learning, etc.

> China launched the world's first quantum satellite last August, to help establish "hack proof" communications, a development the Pentagon has called a "notable advance" [..] The satellite sent quantum keys to ground stations in China between 645 km (400 miles) and 1200 km (745 miles) away at a transmission rate up to 20 orders of magnitude more efficient than an optical fiber [...].
>
> Any attempt to eavesdrop on the quantum channel would introduce detectable disturbances to the system [...]. "Once intercepted or measured, the quantum state of the key will change, and the information being intercepted will self-destruct," Xinhua said.[116]
>
> *Reuters, 2017*

Quantum computers are not yet available for the public use, and only a few organisations provide access to them as a service. Being at the initial stage of their implementation, they are mostly under development by research centres, such as Anyon Systems inc, Cambridge quantum computing limited, D-Wave system corporation, Intel corporation, IBM corporation, Lockheed martin corporation, and are used by technological companies such as Google, Honeywell, IBM, Microsoft.

---

[114] https://www.newscientist.com/question/what-is-a-quantum-computer/.

[115] D. Ventura, T. Martinez, An artificial neuron with quantum mechanical properties, in: Proc. Intl. Conf. Artificial Neural Networks and Genetic Algorithms, 1997, pp. 482–485.

[116] Michael Martina, Chinese quantum satellite sends 'unbreakable' code, Reuters, 10 August 2017, https://www.reuters.com/article/us-china-space-satellite-idUSKBN1AQ0C9.

> Within the decade, Google aims to build a useful, error-corrected quantum computer. This will accelerate solutions for some of the world's most pressing problems, like sustainable energy and reduced emissions to feed the world's growing population, and unlocking new scientific discoveries, like more helpful AI.[117]
>
> *Google, May 2021*

However, the opinions on their potential are not conclusive. Quantum computers by design are not superior in sequential operations and operations with two numbers, and with the present architecture and binary dominance, cannot run binary software faster in any way.

Errors and error propagation in this type of computational devices is still very high and there are no immediate solutions that have been developed. With a small number of qubits (around ten), errors are manageable, but a working quantum computer requires millions of qubits. And the error propagation will be too high to receive accurate results.

Operating systems and software, as well as programming libraries are still at the early stage of development and have not received acknowledgement wide enough to be considered a developed ecosystem with an active community. As software from binary computers cannot be copied to quantum computers, every task would require a new algorithm to be developed.

> Firms like IBM go as far as to claim that quantum computers "will never reign 'supreme' over classical computers, but will rather work in concert with them, since each have their unique strengths,"[118]
>
> *Pednault et al., IBM blog, 2019*

The laws of quantum mechanics are significantly different from the laws of classical mechanics. Similarly, quantum computing is different from classical computing. It is critical to highlight that even at such early stages quantum computers already outperform modern supercomputers in specific tasks. Their **advantages** are feasible in many areas, e.g., resolving issues related to the problem of large integer factorisation, that is virtually unsolvable using conventional computer architecture. Though they are not as fast and powerful as classical computers, and their electronic memory components are a lot slower with a limited number of quantum bits, they can

---

[117] Google, web-site, May 2021, https://blog.google/technology/ai/unveiling-our-new-quantum-ai-campus/.

[118] Edwin Pednault et al. "On Quantum Supremacy", IBM blog, 21 October 2019, https://www.ibm.com/blogs/research/2019/10/on-quantum-supremacy/.

process certain tasks with the reduced level of complexity, as compared to classical computers.

Computational Complexity

Complexity is a computational effort that is required to run an algorithm. It is usually represented as a function *O(n)*. For example, the computational effort of addition is O(n). The effort of adding two numbers increases linearly with the number of steps or items. The computational effort of multiplication is $O(n^2)$. The effort increases by the square of the number of steps or items.

The best algorithm solving the problem of finding the prime factors of an n-digit number (factorisation), is O(e^n^(1/3)). The complexity increases exponentially. Thus, the difference between $O(n^2)$ and O(e^n^(1/3)) complexity is significant. For instance, following this formula, the factorisation of two 800-digit numbers would take several thousand years using a supercomputer.

With the reduced complexity of search algorithms, search tasks in research and commercial applications can be greatly improved. Quantum computers are already used by some of the corporations to run machine learning models to process large volumes of data or, in case of Google, to improve the quality of search results.

Quantum computers can be used for post-quantum encryption, that is meant to be resilient to the attack methods of the quantum era. The quantum computing potential capacity to break all the conventional encryption algorithms has been an issue of concern and conspiracy theories. The concern is that the solution to the prime number factorisation would result in all the internet communications being exposed to the operators of a quantum computer.

Hack-proof communications are a reality

[T] he new demonstration by [Jian-Wei Pan of the University of Science and Technology of China] and his colleagues ensured that Micius would not "know" [the encryption keys]. The trick was to avoid using the satellite as a communications relay. Instead the team relied on it solely for simultaneously transmitting a pair of secret keys to allow two ground stations in China, located more than 1120 kilometres apart, to establish a direct link. "We don't need to trust the satellite," Pan says. "So the satellite can be made by anyone—even by your enemy." Each secret key is one of two strings of entangled photon pairs. The laws of quantum physics dictate that any attempt to spy on such a transmission will unavoidably leave an errorlike footprint that can be easily detected by recipients at either station.[119]

Karen Kwon, Scientific American, 2020

2.4.2 Main Notions

Quantum computing has a set of unique definitions and notions that are required to understand the process of quantum machine learning or quantum computations in general. Those notions include general concepts of quantum mechanics, quantum information, and modern implementations of quantum processing.

There are four principles deriving from the quantum mechanics: *quantum superposition, quantum entanglement, no-cloning* and *no-signaling.* It is worth mentioning, that those phenomena are more relevant to quantum mechanics in general and to underlying principles behind qubits used in quantum computers, rather than to quantum machine learning directly.

Quantum superposition[120] is a principle of quantum mechanics, according to which any two (or more) quantum states can be added together ("superposed") and the result will be another quantum state. The opposite is also true. Every quantum state can be represented as a sum of two or more other states.

The superposition principle reflects the idea of quantum mechanics, where a particle, a group of particles, or a system can be in all possible states at the same time, until that system is measured. In the so called "Copenhagen Interpretation" of the measurement process in quantum mechanics, as formulated by Niels Bohr and Werner Heisenberg in 1925–1927, but not accepted by the whole community of the quantum physicists, including Albert Einstein and Erwin Schroedinger, the wave function describes both a particle (e.g. an electron) and a wave (e.g. a photon of light) before the measurement process.

After the measurement, the wave function "collapses" to a single function, describing a single state. This is a probabilistic non-deterministic process. In quantum computers the basic quantum bit, also called a q-bit or qubit, describes numerous possible combination of 1s and 0s at the same time. The final calculation emerges only once the qubit is measured or "read", and the qubit collapses to either a classical bit 0 or 1 with the same probability. Operating this way a quantum computer allows a "quantum parallelism" with an exponential capacity of registering data in comparison to the classical computers working in parallel.

As qubit falls to one of the basic states that are added together to form the superposition. Those superposed states can result in an interference pattern, that could not have been predicted by the means of classical mechanics.

Quantum interference is a phenomenon of individual particles crossing own trajectories and interfere with own direction.

In quantum computers the interference can be introduced from noise or any environmental events. This interference principle allows the researchers to change the bias of quantum systems for error correction. Similar to reinforcement machine learning,

[119] China Reaches New Milestone in Space-Based Quantum Communications, 25 June 2020, https://www.scientificamerican.com/article/china-reaches-new-milestone-in-space-based-quantum-communications/.

[120] The superposition principle is often considered to be a postulate.

the paths to the "wrong" answers receive "destructive interference" while the paths to the "right" answers get "reinforced".

A quantum algorithm can use superposition to evaluate all possible factors of a number simultaneously. And rather than calculating the result, it uses interference to combine all possible answers in a way that yields a correct answer.

The basic units of quantum information are quantum bits, or *qubits*. For example, a quantum computer, presented by Google in 2019 had 53 qubits, while IBM in 2020 had a 65-qubit computer, with a roadmap to build a 1000-qubit computer by 2023.[121] Qubits can be represented both as waves and as particles. All quantum objects can be observed as waves and particles. This **wave-particle duality principle** allows qubits to interact with each other using the principles of destructive and reinforced interference.

The most serious problem in quantum computing is the decoherence phenomenon, meaning that the qubit can be destroyed by the interaction with the external environment, called noise. Therefore, the robustness indicator of the quantum computer is the maximum time of the coherence, which currently is as little as 100 microseconds.

Quantum entanglement is "the phenomenon whereby a pair of particles are generated in such a way that the individual quantum states of each are indefinite until measured, and the act of measuring one determines the result of measuring the other, even when at a distance from each other."[122] This definition refers to the idea, that two qubits are always in a superposition of two states. Once the first state is measures, the measurements of the second (always opposite) state can be inferred from the first one. Those two opposite qubits are called entangled.

This work of the quantum entanglement is yet to be understood, and even Einstein himself described it as a "spoky action at a distance". Roughly speaking when two particles of the same quantum phenomena are entangled, regardless the spacial distance among the two entangled particles, measuring the state of one particle reveals instantaneously information about the other. The simplest way to understand these quantum phenomena might be that the space of all states in quantum mechanics is not the classical events' space, but an abstract, named Hilbert space, among all possible quantum states having no relations with the classical space of the particle locations.

In quantum computers, changing the state of an entangled qubit will immediately change the state of the paired qubit. The entanglement of qubit is necessary for a quantum algorithm to allow an exponential speed up over classical computer calculation. Quantum superposition and quantum entanglement have alread been

[121] IBM promises 1000-qubit quantum computer—a milestone—by 2023, accessed on 02 May 2021, https://www.sciencemag.org/news/2020/09/ibm-promises-1000-qubit-quantum-computer-milestone-2023.

[122] Oxford Dictionary.

experimentally proven. However the entanglement cannot be used to send information from one particle to the entangled one. Doing this would violate the Einstein principle of relativity that prohibits any phenomenon from being used for faster than light communication. This is known as "no signaling principle". Furthremore the quantum "no-cloning principle" states that it is impossible to clone or copy an unknown quantum state.

Quantum internet

In the researchers' experiment, they investigate a setup containing a length of fibre optic cable with a piece of quantum "memory" at each end. Like the memory in a classical computer, quantum memory is supposed to reliably store information so that it can later be retrieved. The researchers use a particular type of quantum memory that is actually made out of a crystal. When hit by a photon, the crystal vibrates in a distinctive way, different depending on whether the photon is heads versus tails. In this way, the crystal encodes the entanglement information. (In reality, photons do not have two sides like a coin, but more nuanced properties like polarization or spin, which are nevertheless encoded in an analogous way.)

They show with their setup that the photons in each piece of quantum memory maintain entanglement, even after a relatively long time has passed. This means their setup could make for robust nodes in a repeating network. The maintenance of entanglement is indicated by the release of a separate photon, called the "heralding" photon, which is only emitted when the two particles are entangled.

Because their "heralding" photons are generated at standard telecom frequencies, they believe their system can be translated to field-deployed systems and used in real networks between quantum computers, or in other words, a quantum internet.

https://www.nature.com/articles/s41586-021-03481-8

Quantum coherence implies that two waves or particles that are coherent if their qualities (e.g., frequency, waveform, etc.) are identical. Coherence is an ideal scenario. The loss of coherence is called decoherence.

In quantum mechanics particles are described using wave functions, and such phenomena as superposition of multiple qubits can cause quantum decoherence. Those phenomena happen due to quantum interference, natural or artificial, and can result in errors in computation. Thus, to identify where the qubit "is not", the program task has to be rerun over and over again, similarly to training a machine learning model.

To mitigate the risk of quantum decoherence, quantum system has to be environmentally isolated, as any interaction with any other system may result in noise and interference. The temperature of almost absolute zero, at which qubits are kept, allows to isolate the quantum system from any environmental energy.

> The total entropy of an isolated system can never decrease over time, and is constant if and only if all processes are reversible. Isolated systems spontaneously evolve towards thermodynamic equilibrium, the state with maximum entropy.
>
> *Second Law of Thermodynamics*

Quantum computers process information differently, as compared to binary computers, specifically at the level of central processing units. Those problems, that are typically sequential, can be calculated in parallel, or with significantly less steps. For example, a search in an unordered sequence of items will be significantly faster. The complexity will be reduced from $O(N)$ to $O(\sqrt{N})$.[123]

Application of such algorithms, that require significantly less steps is known as **quantum speed-up**.

To conclude, it is worth noting that suspension of qubits in superposition is a complex and sophisticated process, which requires almost absolute zero (1.1 K or negative 272 °C). If the temperature is higher, the memory of a qubit gets corrupted. This causes the issues of scalability, as currently each qubit has a dedicated cooling system. Creating a quantum computer with a million qubits would require equally one million cooling units for each qubit.

Scalability of qubits is challenging as the cooling system is directly connected to each qubit, and there is still no immediate solution to make a joint cooling system for a group of qubits.

As mentioned previously, qubits can be affected by decoherence (i.e., noise) and over time lose required quantum behaviour and stored data. In quantum mechanics no observation can be made without affecting the system and thus the data cannot be extracted without introducing errors. This decoherence is also one of the main reasons a full quantum computer has not been created yet. Furthermore, decoherence significantly reduced the effectiveness of the speed-up that quantum algorithms offer. The logic gates have to be implemented with 0.1% or lower error rates, so that quantum error correction could be feasible.

2.4.3 *Specificity of Quantum Machine Learning*

The emergence of the next-generation computational devices led to the development of the new types of algorithms and quantum machine learning.

[123] Using Grover's algorithm for quantum search.

> [..] a quantum computer's ability to handle optimization and probability problems could significantly improve machine learning and financial trading, as well as solve the thorniest traffic problems. Think of it this way: when all those self-driving hits the roads, quantum computers can help ensure they all take the most efficient routes.[124]
>
> *Wired, 2017*

Quantum machine learning is the integration of quantum algorithms within machine learning.[125] Quantum software uses the architecture of a quantum computer, rather than the conventional binary architecture. The patterns produced by quantum systems are not compatible with binary systems and have to be either revised or expanded.

Quantum interpretation of machine learning models is forecasted to be a lot faster and more precise. Already existing and newly developed algorithms of quantum computing can be used as building blocks to adapt conventional machine learning models to the new architecture, or use it as a stepping stone to move further.

Several quantum-specific machine learning models already exist, such as qubit neuron model[126] or quantum kernel. The machine learning needs data, which is typically collected and stored in a binary format. To be used for quantum or simulated quantum machine learning, binary data has to be converted into quantum data though a set of quantum information processing methods and algorithms.

Reversible/Quantum Machine Learning and Simulation (RQMLS)

Machine learning and artificial intelligence techniques are currently being applied in a diverse number of fields, including molecular simulation, many-body physics, classification, and computational optimization. However, progress in addressing these types of problems is being slowed or stopped when the problem complexity grows exponentially with problem size. Moreover, even when these complexity barriers are overcome, the impact of machine learning solutions are often mitigated by the high energy cost of training and operating the machine learning systems.

In principle, both of these fundamental obstacles—exponentially growing complexity and energy inefficiency—might be overcome using high-coherence

[124] The Race to Sell True Quantum Computers Begins Before They Really Exist, Wired, 6 March 2017, https://www.wired.com/2017/03/race-sell-true-quantum-computers-begins-really-exist/.

[125] Schuld, Maria; Petruccione, Francesco (2018). Supervised Learning with Quantum Computers. Quantum Science and Technology. https://doi.org/10.1007/978-3-319-96424-9.

[126] N. Matsui, M. Takai, H. Nishimura, A network model based on qubit-like neuron corresponding to quantum circuit, Inst. Electron. Inf. Commun. Jpn. III: Fundam. Electron. Sci. 83 (10) (2000), pp. 67–73.

quantum annealers, which are a specific type of quantum computing technology.

DARPA's Reversible/Quantum Machine Learning and Simulation (RQMLS) AIE opportunity aims to:

(1) explore the fundamental limits of reversible quantum annealers;
(2) quantitatively predict the computational utility of these systems for machine learning tasks in simulation, many-body physics, classification, optimization, and other fields; and
(3) design experimental tests for these predictions that can be carried out on small-scale systems.

If successful, these small-scale systems could be scaled to much larger, potentially transformative systems.[127]

Quantum machine learning has **multiple methods**, including quantum neural networks, Hidden Quantum Markov Models, Quantum-enhanced reinforcement leaning, and other methods that are going to be listed in this subchapter. To imitate the established trend in binary methods, many machine learning methods have been simply translated to quantum architecture.

Currently, there exists hybrid classical and quantum models. For example, many Quantum neural networks are the same feed-forward networks that have already been used in binary computers, except instead of neurons they use qubits.

Fully quantum machine learning is the type of machine learning approach where both the model and the hardware architecture are quantum-specific.

Researchers have already analysed the performance of quantum computers for training of the neural networks. While a robust formulation of a neural network is still a fair way away in the quantum realm [25], academics have produced varying methods to represent classical neural networks with quantum circuits. For example, ETH Zurich and IBM Q have already compared classical neural networks and quantum neural networks. Training on Iris dataset, quantum neural network outperformed a classical neural network.

Figure 2.14 shows experimental results of the ETH Zurich and IBM Q, demonstrating how quantum neural network outperforms a non-quantum artificial neural network in trainability and capacity.

Quantum data is usually defined as any data that is present in a quantum system. Quantum data has to be generated using quantum principles of superposition and entanglement, and would require exponentially higher binary storage capacity. Thus, it can be natural or simulated.

Several different methods have already been identified and developed to generate quantum or quantum-compatible data. Those methods include the simulations of the processes, listed in Table 2.7 (Fig. 2.15).

[127] Reversible/Quantum Machine Learning and Simulation (RQMLS), DARPA, https://www.darpa.mil/program/reversible-quantum-machine-learning-and-simulation.

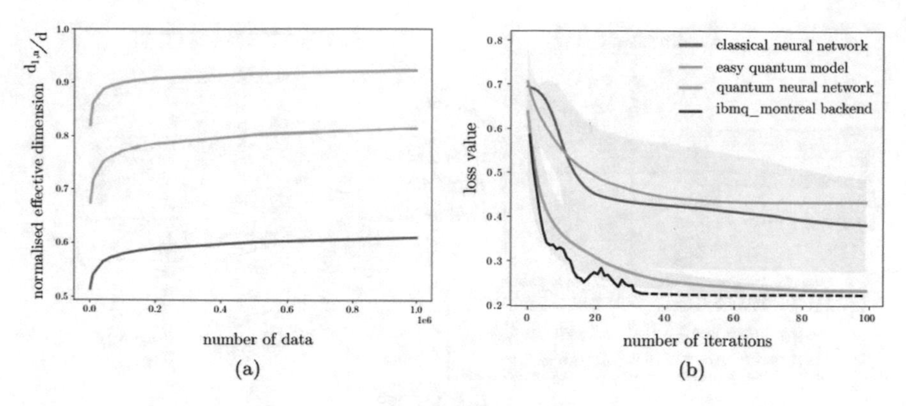

Fig. 2.14 A quantum neural network trained better than a classical neural network [26]

Table 2.7 Quantum-compatible methods of data generation

Method	Applications
Chemical simulation	Material science, computational chemistry, computational biology, and drug discovery
Quantum matter simulation	High temperature superconductivity or other exotic states of matter
Quantum control	Open or closed-loop control, calibration, and error mitigation in quantum devices and quantum processors
Quantum communication networks	Design and construction of structured quantum repeaters, receivers, and purification units
Quantum metrology	High precision measurements such as quantum sensing and quantum imaging

As an alternative approach to denoising the data before its use for machine learning, unsupervised and semi-supervised machine learning approaches can use models that collect features from noisy quantum data. For example, TensorFlow Quantum programming library provides classes and functions, that can be used to train models that disentangle and learn from quantum data using non-quantum computers.

Currently, the **number of** classical **machine learning methods** considerably exceeds the quantum machine learning methods. Furthermore, most of the quantum machine learning methods are the same classical methods, such as Support Vector Machines, Deep neural networks, and principal component analysis, only designed to work with quantum data.

The existing models, that have already been successfully transformed from classical to quantum architecture are:

- Quantum Principal Component Analysis
- Quantum Support Vector Machines
- Quantum Optimization and Quantum Annealing

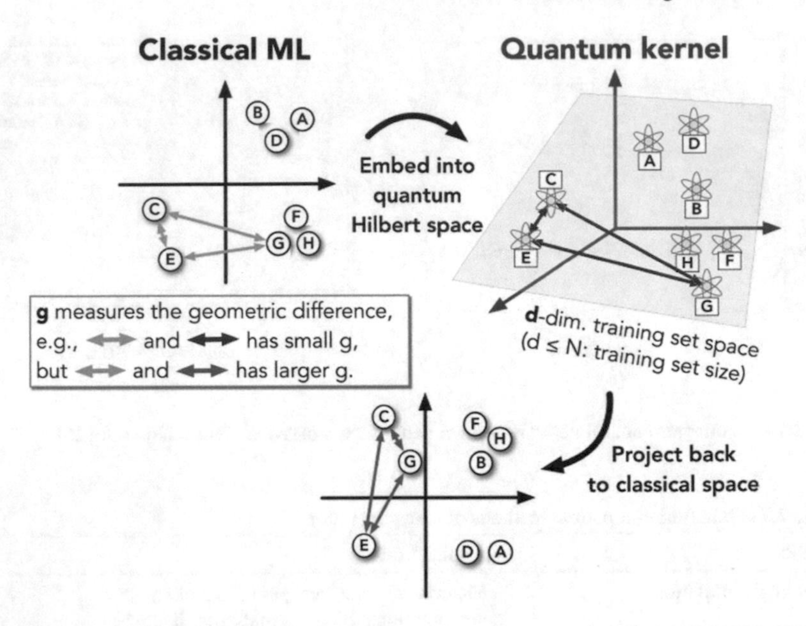

Fig. 2.15 Projected quantum kernel. *Source* https://www.nature.com/articles/s41467-021-22539-9

- Quantum Neural Networks
- Deep Quantum Learning
- Quantum-enhanced reinforcement learning
- Hidden Quantum Markov Models.

In an ideal scenario of quantum machine learning, data, learning model, and the hardware are all quantum, not translated. Several classes of problems could potentially be solved by the fully quantum approach. One of those problems is the discovery, analysis, and replication of the new quantum states. Unsupervised learning and reinforcement learning benefit from a quantum-only version as long as both the method and the information processing devices are quantum.

There is also a challenge in the translation of machine learning algorithms. Replacement of vectors in conventional machine learning with quantum alternatives does not improve the performance as the quantum speed-up is not applied. Thus, the use of a quantum computer is not viable for such algorithms.

A hybrid quantum–classical model can represent and generalise quantum data using classical computers and potentially vice versa. As quantum computers are not sufficiently powerful, a quantum machine learning is not yet sufficiently tested.

However, quantum computers allow to train machine learning models more accurately and faster than binary computers, and combined with the improved search performance, quantum computers are a very promising hardware for machine learning. Quantum computing will also enhance the research capacities in the field of machine learning. With combined applications of machine learning and quantum computing, medicine and pharmacology received the most powerful tool for research, diagnostics, and new drug development. Quantum computers with their unique architecture, can be used for complex simulations, such as chemical simulations, weather forecasting, and other event predictions. In military, they can be used for a variety of tasks, from secure communications to advanced missiles.

In spite of the existing challenges, quantum computers are a promising branch of computational devices for specific applications. The forecast for the development of stable error-free quantum computers is that machine learning will be accelerated by several orders of magnitude. The speed-up offered by quantum operations, capacity to resolve complex issues, enhanced security potential, can justify the cost and complexity of qubit manufacturing.

The above review shows that a potential for the next generation of super powerful and smart computers exists, that will greatly influence the AI development.

The following subsection will summarise the limitations of the current machine learning applications that need to be resolved to move this promising field forward.

2.5 Machine Learning Limitations

Machine learning technology has an extremely promising potential, that is yet to be explored. However, every method has drawbacks. Table 2.8 summarizes the strengths and weaknesses identified through the undertaken literature review conducted during this book writing, and also through the research experiments of the authors.

Currently, a number of limitations of machine learning are considered as barriers on the way to the progress and require additional attention of researchers for their improvement.

Table 2.8 Strengths and weaknesses of machine learning

Strengths	Weaknesses
• Handles large multi-dimensional data • Extracts unknown knowledge from data and identifies trends • Flexible and adjustable • Task does not need to be formulated • Wide applicability, including in detection of unknown cyber threats	• Time and labour consuming for collection of suitable data and its pre-processing • Machine learning bias due to potential errors and limits in data sets • Ambiguous solutions, e.g., multiple answers to the same pattern • Comparatively high error rate • Reliability issues due to lack of clarity in decision-making (e.g., deep learning) • Need in additional efforts for interpretation of results • Need in high computations resources • Cyber security risks due to new types of vulnerabilities • Degrading with time, and need of expert supervision for continuing learning

First of all, machine learning **models can fail**. Among the most basic failures is the training data low quality and its degrading with time. In addition, among the reasons are adversarial attacks (intentional/security failures), but also the technique specific errors or accidents (unintentional/safety failures) (Ram Shankar Siva Kumar et al. 2019).[128,129]

Second, **specific limitations** of the machine learning algorithms, e.g. neural networks, stem from the simplified models of neurons. It is expected that further studies of the human brain work may equally advance the models of neural networks (Gidon et al. 2020).[130]

Things to take into consideration for machine learning improvement

1. "Supervised learning requires too much labelled data and model-free reinforcement learning requires far too many trials. Humans seem to be able to generalize well with far less experience.
2. Current systems are not as robust to changes in distribution as humans, who can quickly adapt to such changes with very few examples.
3. Current deep learning is most successful at perception tasks and generally what are called system 1 tasks. Using deep learning for system 2 tasks that require a deliberate sequence of steps is an exciting area that is still in its infancy."[131]

Yoshua Bengio, Yann Lecun, Geoffrey Hinton (the 2018 Turing Awards recipients for Deep Learning), 2021

In addition, researchers raise concerns over the **machine weaknesses in generalization** when it goes beyond the training data (Bahdanau et al. 2018, 2019).[132,133] The surrounding environment contains such a vast variety of unpredictable relations, connections and sequences, that it is next to impossible to collect data reflecting

[128] Ram Shankar Siva Kumar et al. Failure Modes in Machine Learning, 25 November 2019, at https://arxiv.org/abs/1911.11034.

[129] "Deep Neural Networks Are Easily Fooled: High Confidence Predictions for Unrecognizable Images", Proceedings of the IEEE Conference on Computer Vision and Pattern Recognition, 2015, pp. 427–436.

[130] A. Gidon, T. Zolnik, P. Fidzinski, F. Bolduan, A. Papoutsi, P. Poirazi, M. Holtkamp, I. Vida, and M.E. Larkum. Dendritic action potentials and computation in human layer 2/3 cortical neurons. Science, 367(6473):83–87, 2020.

[131] Yoshua Bengio, Yann Lecun, Geoffrey Hinton. "Turing lecture: Deep Learning for AI", *Communications of the ACM*, July 2021, Vol. 64 No. 7, Pages 58–65. 10.1145/3448250 at https://cacm.acm.org/magazines/2021/7/253464-deep-learning-for-ai/fulltext.

[132] Bahdanau, D., Murty, S., Noukhovitch, M., Nguyen, T., Vries, H., and Courville, A. Systematic generalization: What is required and can it be learned? 2018; arXiv:1811.12889.

[133] Bahdanau, D., de Vries, H., O'Donnell, T., Murty, S., Beaudoin, P., Bengio, Y., and Courville, A. Closure: Assessing systematic generalization of clever models, 2019; arXiv:1912.05783.

the complex world in its entirety. The models require a bigger amount of data for learning, that actually exceeds the information needed for humans as they are able to transfer the knowledge in a better way and interpret new things even if they were not taught (Lake et al. 2017).[134] Reinforcement learning requires even more efforts that the other methods as it is not based on traditional models or datasets.

Currently, one of the ways to improve the quality of machine earning process is scaling it up by generating more data and providing more of the computational power. The improvements are visible, e.g. in chatbots (BlenderBot,[135] Meena[136]), in the new text generators based on deep learning—Generative Pre-trained Transformer (GPT-3) that has 175 billion machine learning parameters as compared with the Microsoft's Turing NLG with a 17 billion parameters, or its predecessor GPT-2 that has 1.5 billion parameters (Brown et al. 2020).[137]

With the scaling up of machine learning parameters, more powerful **computational resources** are required to reduce the training time, as well as the new **hardware architectures** to respond to the needs. Cryptocurrency mining industry has shown the impact of high-complexity computations on the environment. Modernised hardware should allow energy-saving and environmentally friendly computations, which can be achieved through the development of light-weight machine learning methods.

New types of vulnerabilities specific to the machine learning process require enhanced cyber security measures to ensure safety and security at all stages of the AI-system life cycle.

Deep learning insecurities

Deep learning is relatively new and, while a powerful technique, has certain insecurities. Deep neural networks are vulnerable to a form of spoofing attack that uses adversarial data to fool the network into misidentifying false data with high confidence. This vulnerability is prevalent across neural networks in wide use today. While adversarial training can somewhat mitigate these risks, there is currently no known solution to this vulnerability. Additionally, machine learning is vulnerable to "data poisoning" techniques that manipulate the data used to train a machine learning system, thus causing it to learn the wrong thing. Finally, artificial intelligence systems today, including those that do not use deep learning, have a set of safety challenges broadly referred to as "the control problem." Under certain conditions, for example, artificial

[134] Lake, B., Ullman, T., Tenenbaum, J., and Gershman, S. Building machines that learn and think like people. Behavioral and Brain Sciences 40 (2017).

[135] Roller, S., et al. Recipes for building an open domain chatbot, 2020; arXiv: 2004.13637.

[136] Adiwardana, D., Luong, M., So, D., Hall, J., Fiedel, N., Thoppilan, R., Yang, Z., Kulshreshtha, A., Nemade, G., Lu, Y., et al. Towards a human-like open-domain chatbot 2020; arXiv preprint arXiv: 2001.09977.

[137] Brown, T. et al. Language models are few-shot learners, 2020; arXiv:2005.14165.

intelligence tools can learn in unexpected and counterintuitive ways that may not be consistent with their users or designer's intent.[138]

Horowitz, Scharre and Velez-Green, A Stable Nuclear Future? The Impact of Autonomous Systems and Artificial Intelligence, 2019

Limitations related to **security** stem from the lack of agreed safety and development standards for all stages of the AI-system life-cycle. This partially results from the very nature of the machine learning models with non-transparent internal processes, preventing full understanding of the machine learning functionality. The combination of conventional code processing and non-linear logic increases the vulnerability of the complex system. Applying traditional security testing techniques to the machine learning models might not always give a comprehensive insight into the scale of vulnerabilities and their impact. Simply fuzzing a model will not result in the discovery of the machine-learning-related software bugs.

2.6 Conclusion

This chapter has presented an overview of the machine learning landscape and steps, revealed its most essential elements and the way it is implemented in different applications.

It has also covered the advantages and challenges of various types of the pre-deployment machine learning, which can be used for the selection of a more appropriate method for a particular task and a given program.

With the growing availability of programming libraries and learning literature, virtually anyone can engage into the development of beginner-level AI-systems. However, as any new technology, machine learning is not yet fully explored and not perfect. Entrusting it with mission critical systems is premature. Understanding and adopting machine learning without false expectations would allow researchers and engineers to use it with maximum efficiency, and build on it, allowing technological progress to continue.

To be effective during their full life-cycle, the trained systems should continue learning based on the newly generated data. However, the post-deployment learning, especially in a complex environment, is an expensive process requiring high level expertise, robust hardware capacities for data collection and procession. More knowledge is needed in the areas of more efficient learning, optimised hardware and software requirements during and post-training, learning without forgetting, and continuous learning without introducing errors.

---

[138] Michael C. Horowitz, Paul Scharre and Alexander Velez-Green, A Stable Nuclear Future? The Impact of Autonomous Systems and Artificial Intelligence, working paper, December 2019, pp. 5–6, https://arxiv.org/ftp/arxiv/papers/1912/1912.05291.pdf.

The ongoing research in machine learning and real-life events, shows that the trained systems not only keep making errors, but are also highly vulnerable to increasingly sophisticated cyber attacks. Moreover, machine learning itself can be used for generating AI-empowered cyber arms. Thus, the next Chapter will review a specific practical application of machine learning in defence against cyber attacks.

References

1. Abaimov S, Martellini M (2020) Cyber arms: security in cyberspace. CRC Press
2. Poole D, Mackworth A, Goebel R (1998) Computational intelligence. Oxford
3. Kurzweil R (2005) The singularity is near: when humans transcend biology. Viking Penguin, New York
4. Feigenbaum E, Feldman J (eds) (1963) Computers and thought
5. Nilsson N (1965) Learning machines. McGraw Hill
6. Duda R, Hart P (1973) Pattern recognition and scene analysis. Wiley Interscience
7. Stuart R, Peter N (2003) [1995] Artificial intelligence: a modern approach, 2nd ed. Prentice Hall
8. Nilsson N, Kaufman M (1998) Artificial intelligence: a new synthesis. Morgan Kaufmann Publishers Inc
9. Valiant LG (2013) Probably approximately correct. Basic Books
10. Mitchell T (1997) Machine learning. McGraw Hill, New York
11. Hebb D (1949) The organization of behavior: a neuropsychological theory. John Wiley and Sons, New York
12. Cangelosi A, Fischer MH (2015) Embodied intelligence. Springer
13. Alpaydin E (2020) Introduction to machine learning, Fourth ed. MIT
14. Asimov I (1942) Runaround, astounding science fiction (March 1942)
15. Sutton RS, Barto AG (2018) Reinforcement learning: an introduction. MIT Press
16. Stuart R, Peter N (2009) Artificial intelligence: a modern approach, 3rd ed. Prentice Hall Press
17. Vapnik V (2013) The nature of statistical learning theory. Springer Science and Business Media
18. Goodfellow I, Bengio Y, Courville A (2016) Deep learning. MIT Press
19. Jordan MI, Bishop CM (2004) Neural networks. In: Tucker AB (ed) Computer science handbook, Second edition (Section VII: Intelligent Systems). Chapman & Hall/CRC Press LLC, Boca Raton, Florida
20. Efron B, Tibshirani RJ (1994) An introduction to the bootstrap. CRC Press
21. Watkins CJCH, Dayan P (1992) Q-learning. Machine learning **8**(3–4): 279–292
22. Mnih V, Kavukcuoglu K, Silver D, Rusu AA, Veness J, Bellemare MG, Graves A, Riedmiller M, Fidjeland AK, Ostrovski G, Petersen S (2015) Human-level control through deep reinforcement learning. Nature, **518**(7540): 529–533
23. Schulman J, Levine S, Abbeel P, Jordan M, Moritz P (2015, June) Trust region policy optimization. In: International conference on machine learning (1889–1897). PMLR
24. Feynman RP, Kleinert H (1986) Effective classical partition functions. Physical Review A **34**(6), 5080
25. Schuld M, Sinayskiy I, Petruccione F (2014) The quest for a quantum neural network. Quantum Information Processing **13**(11): 2567–2586
26. Abbas A, Sutter D, Zoufal C, Lucchi A, Figalli A, Woerner S (2020) The power of quantum neural networks

Chapter 3
Defence

In cyber security, defence is an essential and the hardest part. As is well known, to protect the whole system it is necessary to know all its vulnerabilities, while an attacker may need only one to breach it. Thus, defenders by default are in a disadvantageous position.

More and more devices become controlled by computerised systems. Cyberspace is growing and impacts the physical world through increased information flows. With information and communication technology developing at a skyrocketing pace, previously small and isolated computer networks are becoming more interconnected and accessible through the internet. Those networks are used in industries, business, research, entertainment, and other area of our lives. Design and deployment of reliable network systems and connection between them with higher level of security is an important task for engineers and developers.

Wireless communication, direct interaction between machines, collection and procession of large volumes of data, mutational capacity of cyber arms enhanced with learning potential are strong risk factors, bringing new types of threats. The attack surface expands proportionally to the number of internet-connected systems, entry points into the network. Furthermore, malicious actors are developing increasingly sophisticated types of attacks, new zero-day exploits and malware that evade security measures. In addition to defending against external threats, defenders must also guard against insider threats from individuals or entities within an organisation that misuse their authorised access.

Artificial Intelligence (AI), outperforming humans in the reaction speed, volume of data procession, round the clock operation mode, is **a new evolutionary instrument in cyber security** with an increasing potential. **Machine learning for defence** implies the use of machine learning models and algorithms for code analysis, vulnerability detection, computer state analysis, hardware and software behaviour analysis, and other techniques, that help to combat any type of cyber-attacks. Intelligent

S. Abaimov and M. Martellini, *Machine Learning for Cyber Agents*,
Advanced Sciences and Technologies for Security Applications,
https://doi.org/10.1007/978-3-030-91585-8_3

models do proactive scanning of potential vulnerabilities, verification and validation of defence mechanisms. Without interacting with the perimeter defences, these advisory systems can aid human operators for further decisions and actions.

The AI-enhanced Intrusion Detection Systems (IDS) are finding a wider application, with the machine learning algorithms being more responsive to the defence needs. Their advantage is not only in taking non-linear decisions faster than humans, but also in their capacity to identify yet unknown threats and protect from them. Growth of unknown attacks and changing of attack vectors shifted the defence mentality from routine procedures and orientation to specific attacks, to enhancing the whole "immune system". For example, machine learning is used by Darktrace "to define what "normal" looks like for any network and all its devices and then report on deviations and anomalies in real time".[1]

Having a highly secure network means having a zero trust approach to security devices, the assumption that threats can originate from anywhere and any device. Zero Trust Architecture is a cyber security model based on maintaining strict access controls for any connected entity.

A Zero Trust view of a Network[2]

(1) The entire enterprise private network is not considered an implicit trust zone
(2) Devices on the network may not be owned or configurable by the enterprise
(3) No resource is inherently trusted
(4) Not all enterprise resources are on enterprise-owned infrastructure
(5) Remote enterprise subjects and assets cannot fully trust their local network connection
(6) Assets and workflows moving between enterprise and non-enterprise infrastructure should have a consistent security policy and posture.

A typical network would have much lower levels of security, than a zero trust architecture. Availability of Security-as-a-Service and cloud security has already resulted in the improved security posture, but the networks of small and medium-size businesses unable to bear heavy costs of expertise and equipment still remain vulnerable.

---

[1] Scott Rosenberg, "Firewalls Don't Stop Hackers, AI Might," Wired, August 27, 2017, https://www.wired.com/story/firewalls-dont-stop-hackers-ai-might/.

[2] Zero Trust Architecture, NIST SP 800–207, August 2020, https://nvlpubs.nist.gov/nistpubs/SpecialPublications/NIST.SP.800-207.pdf.

How AI-enhanced defence works

Conventional cybersecurity tools look for historical matches to known malicious code, so hackers only have to modify small portions of that code to circumvent the defence. AI-enabled tools, on the other hand, can be trained to detect anomalies in broader patterns of network activity, thus presenting a more comprehensive and dynamic barrier to attack.[3]

However, we should always have in mind that inheriting the same dual use qualities from any cyber arms, these smart "cyber agents" can equally serve the offence purposes, feeding a growing number and variety of attacks as well as generating new attack vectors. Among them are remote code execution, denial of service and phishing attacks, that will be discussed later in this book. Security of the AI-enhanced cyberspace is now a more difficult task and innovative solutions are needed.

Offense and defence with the same algorithm

DARPA's 2016 Cyber Grand Challenge demonstrated the potential power of AI-enabled cyber tools. The competition challenged participants to develop AI algorithms that could autonomously "detect, evaluate, and patch software vulnerabilities before [competing teams] have a chance to exploit them"—all within a matter of seconds, rather than the usual months.[4] The challenge demonstrated not only the potential speed of AI-enabled cyber tools but also the potential ability of a singular algorithm to play offense and defence simultaneously. These capabilities could provide a distinct advantage in future cyber operations.[5]

DARPA, 2016

The below subchapters will cover the machine learning application in cyber security in more detail. For a more structured representation, cyber security will be subdivided into computer and network security, though in some cases it may not be that straightforward and can include both.

---

[3] Austine Groves, Paul Schulte, The race for 5G Supremacy. Why China is surging, Where Millennials Struggle, and How America can Prevail, World scientific publishing Co., 2020, p. 220.

[4] 'Mayhem' Declared Preliminary Winner of Historic Cyber Grand Challenge," August 4, 2016, https://www.darpa.mil/news-events/2016-08-04.

[5] Daniel S. Hoadley, updated by Kelley M. Sayler, "Artificial Intelligence and National Security", 10 November 2020, Congressional Research Service Report, United States, p. 2, https://fas.org/sgp/crs/natsec/R45178.pdf.

3.1 Machine Learning for Cyber Security

Cyber security, in its broad meaning, is the collection of policies, techniques, technologies, and processes that work together to protect confidentiality, integrity, and availability of computing resources, networks, software programs, and data from attacks. With AI-enhanced evasive cyber agents, cyberspace is becoming a space without trust, where every device and an application can be potential exploited and cause a threat. Constantly evolving and mutating attack tools can no longer be effectively detected using fixed and linear defence methods. This equally requires strengthening the **cyber threat intelligence**, protective measures covering not only operational, but also tactical and strategic procedures and policies in any organization.

At the level of corporations and industrial facilities, machine learning can be used in **strengthening security policies** that define network rules and controls, divide users into groups and networks into zones, restricting or allowing specific communications between users and applications. The AI-systems can process the incoming traffic and compare it against security policies. Currently, most commonly used security policies are static and created by human operators. Using of manual methods in creation and automatic generation of policies results in duplication and redundancy and increased complexity. Improving security policy development through machine learning will enhance protection and stimulate further research in AI for security.

Proactive defence measures and risk mitigation have become an essential condition for non-interrupted smooth business processes. Cyber defence mechanisms enhanced with machine learning exist at the application, network, host, and data level. They can be of general use, detecting and preventing automated activity and not skilled attackers, or customised, responding to specific needs, e.g. in critical infrastructure, nuclear weapons defence, etc. There is a wide variety of tools—such as firewalls, antivirus software, web application antivirus, intrusion detection systems, and intrusion protection systems—that were designed to prevent attacks and detect security breaches.

The most popular research areas in the use of machine learning for cyber security include network and host intrusion detection, malware analysis and classification, code analysis and vulnerability detection, and botnet traffic identification.

Machine learning application has revolutionised cyber security, through providing flexible and adaptive means to solving complex security problems, that typically require human operators and are not easily solvable by linear algorithms. For example, filtering spam emails or traffic analysis seems impossible for being implemented by human operators at large scale. Machine learning-based systems are designed in such a way that they can potentially predict the early signs and prevent the attack before it reaches the stage of impact.

Cyber attacks implementation follows a fixed sequence of steps. There are several methodologies and frameworks to conduct offensive security assessment. One of them, exclusively related to real-life cyber attacks, is known as Cyber Kill Chain[6]—a model commonly used to analyse and describe the malware attack behaviour. Knowledge of those steps allows researchers to break attack behaviour into parts, examine each one of them and develop mitigation measures with the use of machine learning techniques.

Any attack before succeeding at delivering final payload, shows signs of compromise and breaches, manifests infiltration attempt. For example, with the datasets and existing methodology, a trained model can potentially detect a large-scale attack at *reconnaissance* stage, before the malicious actors manage to infiltrate the network. Another example is using machine learning for network risk assessment, to evaluate and rate the most vulnerable devices or groups of devices in a network against the most valuable resources in the same network or supply chain.

Currently, cyber security is the responsibility of trained technical experts, such as IT support staff, system and network administrators, Chief Information Security Officers, etc. Though being experts, a human factor is present in their decisions that may be influenced by emotions, fatigue, oversight, negligence, etc. The AI-empowered technologies are never influenced by these human weaknesses, take decisions much faster and process bigger volumes of information. Also, human operators physically cannot respond and counter the influx of threats increasing with the unprecedent speed. Round the clock efficiency of machine-learning-based systems can allow to **reduce the workload for experts** by carrying out real-time analysis and addressing at least some of the alerts that would have been previous handled by people. These can be single models, or chains of multiple decision-making systems that considerably increase the performance of conventional intrusion detection systems and reduce the human resources needed for cyber security.

Machine learning can also help to **enhance indirect security-related tasks** that also improve the efficiency of human operators. It can process the collected data, analyse, scale, prioritise and visualise in an optimal way, so that a human operator can make a final "yes or no" decision in cases that cannot be made autonomously. Machine learning prediction capacities bring to their attention the trends in cyber attacks, thus contributing to a safer cyber space in general.

Probably one of the simplest ways to effectively apply machine learning for real-time defence is **spam filtration**. Conventional filtering methods dissect emails into components, while machine learning can use natural language processing to analyse if the email is malicious or not. Even though it may seem simple, phishing and spam emails are the preferred entry point into corporate networks and malware distribution method by the attackers, as it exploits the human factor. As spam and phishing methods become increasingly more complex and evasive, and thus more difficult to detect, it is important that machine learning gets wider acceptance in this particular area of cyber security.

Malware analysis is a complex task that requires extensive human attention. Modern malware can autonomously replicate with changes, to avoid detection while

[6] The Cyber Kill Chain, Lockheed Martin, https://www.lockheedmartin.com/en-us/capabilities/cyber/cyber-kill-chain.html.

still delivering the same impact. This capacity makes malware invisible to the signature-based analysis. Machine learning can be used for the analysis of such malware transformations and detect this malware by signs of activities, rather than code signatures.

As **supply chains** being increasingly adopted and, in some way, globalised, the manufacturing of electronic devices is no longer limited to a single town or even a country. Resources, intellectual property, assembly and logistics are now connected worldwide. As a part of such processes, third-party suppliers are involved in design and assembly of software and hardware. With the growing attention of both defenders and attackers to supply chain, it is now considered as highly susceptible to a range of threats. These threats that may affect government, transportation, medical sector, military, and other critical infrastructures. Machine learning can be applied for the analysis of the supply chain software and hardware components, and enhance its security.

Supply chain attacks

According to the new ENISA report—Threat Landscape for Supply Chain Attacks, which analysed 24 recent attacks, strong security protection is no longer enough for organisations when attackers have already shifted their attention to suppliers. This is evidenced by the increasing impact of these attacks such as downtime of systems, monetary loss and reputational damage. Supply chain attacks are now expected to multiply by 4 in 2021 compared to last year. Such new trend stresses the need for policymakers and the cybersecurity community to act now. This is why novel protective measures to prevent and respond to potential supply chain attacks in the future while mitigating their impact need to be introduced urgently.[7]

With all the benefits of machine learning, there come **drawbacks and costs** to using this new technology for defence. Computers have high-speed flow of data between internal components and between applications. In **computer security**, machine learning has not yet reached enough optimisation sufficient for real-time processing of every operation and system call. Similarly, in **network security**, machine learning introduces too many computational steps to verify every network packet transferred between machines. Data flow in modern networks is too high for a machine learning to analyse it in real time in a sufficient depth without slowing down the network. As a solution, network traffic flow analysis is used to detect anomalies, fluctuations, and spikes in network activity, but detection of anomalies yields a lot of false positives from constantly changing activities of employees, users, and regularly updated software.

[7] European Union Agency for Cybersecurity, https://www.enisa.europa.eu/news/enisa-news/understanding-the-increase-in-supply-chain-security-attacks.

Machine learning cybersecurity applications typically extract low-level features from the network sessions that are already complete (have the beginning and the end of the session recorded), which prevents those experimental systems from detecting the attacks in real life.

At the current, research-focused stage, models can recognise patterns in the carefully tailored features, but only in the general context, and as a sequence of numbers. To make the model learn with a specific context, the dataset should contain features of temporal and spatial nature, as well as the relationships between events or alerts. In addition, many types of network traffic sessions can be both malicious and non-malicious at the same time, based on the context. For example, network discovery is a communication process that is required for the network to operate without malfunctions, so the packets can be delivered from one device to another. The same network discovery can be used by attackers for network enumeration and reconnaissance.

The **accuracy** in intrusion detection is **not sufficient** to be fully reliable. Machine learning-based systems have a high number of false positive alerts and a noticeable number of false negatives. Intrusion detection and anti-virus systems generate a high number of **security alerts**, especially in larger organisations. Even with 99.99% accuracy of the intrusion detection, 1 in 10,000 network sessions can be flagged as malicious, which would happen every few seconds in a medium-sized network. An IDS in a real-life scenario would generate millions of alerts per day. Those alerts maybe be about automatic tools scanning, failed attack attempts, or simply false alarms of non-malicious events. Human operators can get exhausted by these alarms. To address this challenge, a higher-level semi-autonomous management system usually analyses the alerts and prioritises the display of those that need human attention.

Machine learning requires at a minimum a computer, an operating system, a compiler, and a few programming libraries to operate. The introduction of machine learning increases the complexity of the final system, and introduces new points of failure and vulnerabilities.

Designed to speed up the work efficiency, interdependencies and complexities of industrial and corporate wireless systems, **machine learning** generates multiple cyber security **vulnerabilities**. The strongest advantage of machine learning, to collect and process big amounts of data, can be its biggest weakness. Data can be compromised, infected with malicious inputs, it can also be leaked or poisoned by adversaries.

Data poisoning

In the machine learning community dataset poisoning is a popular example of attacks against machine-learning-based systems.

Backdoor attack implies the injection of the compromised samples into the memory of the model, while clean-label poisoning attack causes misclassification without direct manipulation of the labelled data. Clean-label poisoning attack causes misclassification for a specific test sample, without affecting the performance of the model for any other instances.

The access to wireless devices provides immediate penetration to internal networks, and in highly critical networks even the lowest unauthorized privileged access can compromise the mission. Increasing sophistication, accidental or intentional misconfigurations of equipment and exponentially growing number of vulnerabilities require extensive monitoring, reinforced assessment and testing of the wireless equipment and software.

Most vulnerable devices

In 2017, Darktrace saw a 400% increase in the number of IoT (Internet of Things) security incidents in its customers' networks.

Connected objects are among the most vulnerable devices in the IT world. Half a million IoT devices were vulnerable to a botnet named Mirai, which launched mass-scale denial of service attacks in 2016, the largest attack in history at that time. Darktrace has intercepted attackers that have targeted the biometric scanners used to keep unauthorized personnel out of manufacturing plants, internet-enabled drawing pads used by architects, and even a large fish tank that automatically dispenses fish food, but whose internet connection was exploited to exfiltrate data out of a large entertainment corporation.

Catching the Silent Attacker, and the Next Phase of Cyber AI, Darktrace Threat Report, 2018, p. 2

Human operators interact with remote computers through web applications that in their turn more often than not contain numerous vulnerabilities. These vulnerabilities that come with design or through improperly deployed configuration and updates, remain unnoticed due to the lack of time for a proper testing at the end of the development cycle, or high costs of the adequate testing tools.

Web applications

Web applications are the most common way of breaching the system. Specifically in web applications, machine learning is used for the analysis of input attempts, inputs and behaviours, generated by users and potential attackers. Input data cannot be analysed on the client side, and has to be processed at the server side of the application. Given that in this case the malicious payload will eventually reach the server, the intrusion detection system also has to be protected against those inputs.

Web applications can also be monitored for malicious behaviour, in order to minimise processed data and delays, and to maximise the performance of the IDS. Behaviour can be analysed without processing the payload at all, and only by using features of metadata as input into the classifier. Those can include events like:

- Repeated attempts to access search pages with zero results (might hint on fuzzing);
- Repeated attempts to pages that do not exist (directory bruteforce);
- Repeated attempts to the same page with false parameters (injection attempts);
- Repeated changes of personal settings (manual hacking attempts).

Any machine learning based security system would benefit from operating under the guiding principles based on data-driven decisions. The advanced data analysis solutions, powered by machine learning that relies on previously recorded and real-time data, can mitigate the challenges and provide automated and intelligent response. Existing dataset and constantly emerging new machine learning methods can be used to create new robust network protocols, that would boost the security of communications and make it resilient to a range of future attacks.

Machine learning, as any information technology, has a dual-use nature. It aids security professionals but can easily be misused by the malicious actors and creates potential threats. Similar to defenders, attackers keep developing and improving the arsenal of attack tools and vulnerabilities. As the sophistication of the attack methods keeps growing, machine learning models can cooperate with human operators in identifying the threats and detecting the attacks.

3.2 IDS Supporting Human Operators

An ideal intrusion detection system (IDS) should combine in itself both computer and network security approaches. Which implies that the IDS has to be deployed in such a way, that it can simultaneously detect the network activity and the activity at the application level of the operating system.

Conventional non-machine-learning IDSs are widely used in commercial and industrial networks. They are automated, heavily reliant on signature-based detection, and optimised in performance. However, those systems can only detect the exact matches or triggers. Many alerts in such systems get recorded to the event logs without being displayed to the administrators, some of the attacks may stay unnoticed for weeks or even months.

IDS trained with machine learning techniques, have already established their presence in corporate and industrial networks. Those systems monitor network activity and generate alerts when anything suspicious happens in the entire network or on a

Table 3.1 Comparison table of conventional and machine learning IDS

Functions	Conventional IDS	Machine Learning IDS
Detection method	Hard-coded "Signature"	"Statistical" measurement
Detects	Only known attacks	Deviations from "normal"
Accuracy	Accurate only for matching signatures	Can detect non-typical behaviour and potentially unknown attacks
Update	New signatures	May not need any update
Complexity of deployment	Designed for consumer-grade operating systems	Experimental and sensitive to versions of required libraries
False alerts	Low	High

specific device. Unlike conventional IDS, they have an innovative approach to cyber attack detection and the means to potentially execute it, as presented in Table 3.1.

Based on their action, IDS can be active and passive. **Active IDS**, also known as an intrusion detection and prevention system, generates alerts and logs entries along with commands to change the configuration to protect the network. **Passive IDS** detects malicious activity and generates an alert or logs without taking any action.

Using the deployment criteria, IDS can be subclassified as **host-based-IDS** (HIDS) or **network-IDS** (NIDS). HIDS is deployed on the single information host, one specific device. Its task is to monitor all the activities on this single host and scans for its security policy violations and suspicious activities. The main drawback is the need to deploy it on all the hosts (computers) that require intrusion protection, which results in extra-processing overhead for each node and ultimately degrades the performance of the IDS. In contrast, NIDS is deployed on the network with the aim to protect all devices and the entire network from intrusions. This system will constantly monitor the network traffic and scans for potential security breaches and violations.

NIDS can be deployed on a computer, on a server, or on a communication link (a bridge or switch). If designed with machine learning, it will need to use programming libraries for the operation of the model or models, and will require an operating system. Though most of modern communication links tend to be low or medium performance computers, they nevertheless satisfy the architectural and hardware requirements. For low-power devices, every step in the intrusion detection should be optimised for maximum performance, to reduce the delays in traffic analysis.

Similarly, HIDS might also reduce the performance of the device, but at a different level. As NIDS slows down the communication between the devices, the HIDS will increase the time required for some processes to perform designated tasks, slowing down overall workflow.

Based on their approach to attacks detection, IDS can be anomaly based or misuse based. **Anomaly IDS** is implemented to detect attacks based on the recorded "nominal" behaviour and deviations from it. Therefore, it compares the current real time

Table 3.2 Limitations of conventional IDS

Criteria	Risks and limitations
Response to attacks	Only known attacks get detected. This causes sophisticated and complex threats to go often unnoticed
Network interference	Network "noise" can significantly reduce the capabilities of the IDS by generating a high false-alarm rate
Software updates	Constant software updates are required for signature-based IDS to keep up with the new threats
	IDS monitor the whole network, so are vulnerable to the same attacks the network's hosts are. Protocol-based attacks can cause the IDS to fail
Potential	Network IDS can only detect network anomalies which limit the variety of attacks it can discover
	Network IDS can create a bottleneck as all the inbound and outbound traffic passes through it
	Host IDS rely on audit logs, any attack modifying audit logs threaten the integrity of HIDS

traffics with previously recorded normal real time traffics. This type of IDS is widely used as it has the ability to detect new types of intrusions. At the same time, it yields the highest number of false positive alarms, which means there is a large number of normal packets or sessions falsely identified as attacks. **Misuse-based IDS**, as oppose to anomaly-based, learns about malicious activity first, and only then operates over a network with non-malicious traffic. This problem can be better solved by machine learning models, than by conventional IDS due to their limitations. Some of those limitations are presented in Table 3.2.

The machine-learning-based IDSs are not yet wide-spread, being used mostly for research, and in lesser extent—by start-ups. Typically, an experimental IDS would have machine learning algorithms (e.g., Random Forest, Random Tree, Decision Table, DNN, Naive Bayes, and Bayes Network), and it will be evaluated for accuracy using date sets, such as Knowledge Discovery in Databases (KDD) which includes several different types of attacks (Denial of service, remote code execution, and privilege escalation).

Attacks detection in the literature is typically considered as classification problem because the target is to clarify whether the packet is normal or whether it is an attack packet, and if so—identify its type.

As any technological inventions, machine-learning-based IDSs have their benefits and challenges. They have flexible configuration, adaptive to networks they are deployed in, capable in detecting anomalies. However, they slow down the network, demonstrate high rate of false positives, require large volumes of data for training, can learn false solutions, need complex software and hardware, as well as the supervision of human operators.

3.3 Network Security

Network security focuses on the traffic and traffic flow analysis, and network hardware and software behaviour, as well as on the behaviour of network communication links (ports).

There exist non-machine learning solutions, e.g. Deep packet inspection (DPI) is one of the techniques used for a more detailed analysis of packet headers and payloads (transferred data) predominantly in high-security or industrial networks. Multiple machine learning-based implementations for the network security have been developed over the past few decades. Supervised machine learning can be used more effectively for a deeper analysis, while the unsupervised one is more efficient in anomaly detection.

In practice, a combination of both methods of learning is used for the development of the network immune systems. To avoid the slowdown, the innovative intrusion detection methods aim to detect anomalies and cyber attacks by analysing the traffic flow (and metadata), rather than the contents of the packets.

Packets (instead of segments or frames) are a basic unit of data transfer over the network. These packets contain details about the sender and receiver of the packet, protocol details, and data that is being transferred. The part that contains everything except the data is called a *header*. The *data* in the packet is the payload from the application layer protocols (Fig. 3.1).

Packets were selected to be used for the intrusion detection analysis as they contain communication data (payload) and sufficient metadata in the *header* about the origin, destination, and mode of delivery of the packet. Packets can also be processed as they are received, without saving them to storage memory. Those properties can allow real-time detection of ongoing attacks, as well as identification of the data origin. However, a single packet does not reflect the full behaviour in the network

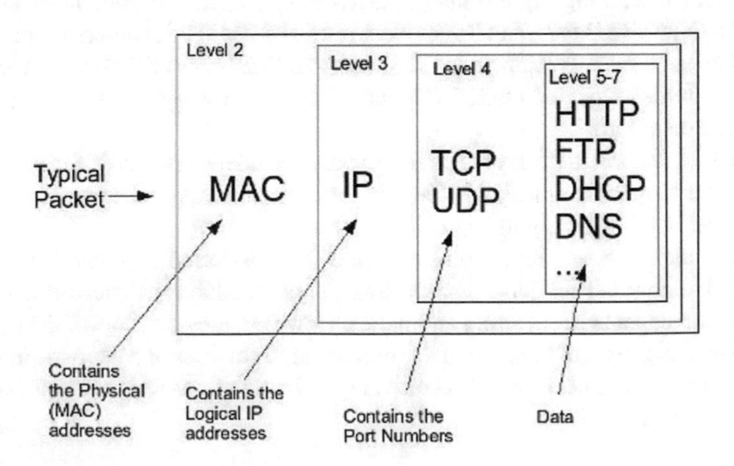

Fig. 3.1 Example of a network packet

and typically cannot describe the entire attack. For example, a single packet from one device cannot describe a DDoS attack, that usually implies thousands of devices sending thousands of packets per second. Analysing multiple packets or even entire traffic sessions at once will still result in "per packet" processing, and a method like this will increase the computational requirements even further.

Traffic analysis at a packet level can be conducted using two methods: *packet parsing* and *payload analysis.*

3.3.1 Packet Parsing-Based Detection

At the network level of communication there is a number of established protocols, most common of which are HTTP/HTTPS, DNS, FTP, and SSH. All those protocols are different in the format and the analysis of their *headers* is used instead. The *headers* can be extracted using either prebuilt tools (e.g. Wireshark) or using programming libraries. Once the *headers* are extracted, their values can be treated as features for machine learning models, supervised or unsupervised.

Multiple clustering methods have already been applied for this particular approach to intrusion detection and have proven that unsupervised learning allows to analyse traffic headers without payload and still have a 99% detection rate for traffic analysis, as long as the headers are not encrypted. The method currently yields high rate of false positives, but the research is ongoing with promising results.

3.3.2 Payload Analysis-Based Detection

When the *header* analysis is insufficient, the *packet payload* is analysed. This type of analysis avoids protocol dependency related to headers and uses only application-level data. This type of data can be pre-processed for machine learning models using a variety of purpose-specific pre-processing methods, including natural language processing. The encrypted payloads introduce new complexity, as they have to be decrypted. The whole process is fairly labour-intensive and compromises privacy of the communication.

This type of data can be analysed by any machine learning method. However, out of all machine learning methods, deep learning is very effective at processing such data. Deep learning models can effectively train to 99% accuracy to detect malicious activity just by analysing payloads. CNN, LSTM, and autoencoders have all shown accurate results. Unsupervised learning approaches usually reach only 98% which is still sufficient to consider unsupervised learning as an alternative to the supervised methods.

Adversarial machine learning was used to create additional synthetic samples to already existing malware datasets, with an attempt to create malware traffic that is as close to non-malicious traffic as possible. For example, if the traffic recordings of social network activity are taken, malware can blend in into a corporate network, while based on ICS traffic, malware can hide from intrusion detection systems in the industrial networks.

Specific data, such as text, code, or queries, is analysed using methods arguably similar to the natural language processing. It is processed in the same way, as conversational robots and programs process and classify text.

As an alternative to the traffic analysis, the traffic flow can be analysed instead. This approach allows the analysis of multiple packets, their directions and size, without looking at the payloads. Such an approach would reduce computational requirements and improve privacy.

Traffic flow specifically includes the analysis of traffic streams and can be active or passive. Typically, the passive analysis used in corporate networks, as the IDS has administrator-level access to network and does not require active participation in the interrogation of each device.

3.4 Computer Security

Computer security is the study of the security of a specific system interacting within its own bounds. It includes hardware, firmware, operating system, processes, directly connected (but not networked) devices, and other events that happen inside the computer itself. This also includes exploitation of the vulnerabilities that are "local". For example, privilege escalation through kernel-level exploit.

Machine learning in computer security can be applied for the analysis of:

- hardware behaviour
- operating system behaviour
- software behaviour
- user behaviour
- specific application behaviour analysis
- malware analysis.

Levels of defence analysed by machine learning include the

- state of the system or an application
- behaviour: change of states in the system, software, or data
- data flow: analysis of metadata
- specific data: unique cases that do not fit into previous categories.

State of the system or an application implies the set of all the possible configurations and setting in the system. Thus, behaviour of the system is the change of states of that system, software, or stored data. Those can be changes in data, configurations, or time between certain actions. As oppose to system behaviour, network behaviour is—network traffic flow and data flow analysis is the analysis of metadata.

Specific data include numeric or alpha-numeric identifiers, that are being transferred through traffic, as well as information about the network that is not included into the network traffic flow.

3.4.1 Hardware Behaviour

Changes and fluctuations in typical behaviour of hardware can be monitored through the direct input from the operating system, or from programming libraries, used for machine learning deployment. The abnormal behaviour of hardware can potentially reveal the presence of malware. Thus, changes in voltage, heat, and load on different hardware components of the system can hint on malicious activity. For example, overheating in systems that work with average load could mean the presence of a malicious process, encrypting the data or mining cryptocurrency.

Specifically, firmware can contain malware that may not be detected by the operating system analysis. For example, Central Processing Unit (CPU) of a modern computer with its own operating system (i.e., microkernel) may be injected with malware. Firmware and microkernels are more secure due to their robustness by design, but it does not make them invulnerable to cyber attacks. Higher usage of CPU or unintended autonomous behaviour from software are all signs of malfunction in hardware.

Even though hardware itself is usually read-only and isolated from the reach of the operating system, many hardware modules require direct input. To protect the hardware, there exist various approaches, such as physical input detection, monitoring of hardware behaviour, regular reset of firmware, and supply chain assurance.

Hardware Trojans

Hardware Trojans are malicious hardware inclusions that leak secret information, degrade the performance of the system, or cause denial-of-service. The risk of Hardware Trojans is increasing due to the outsourcing of the very large-scale manufacturing process to different third-party entities, which results in "supply chain" attacks.

Below is the explanation of how a hardware trojan can be detected using intrusion detection systems.

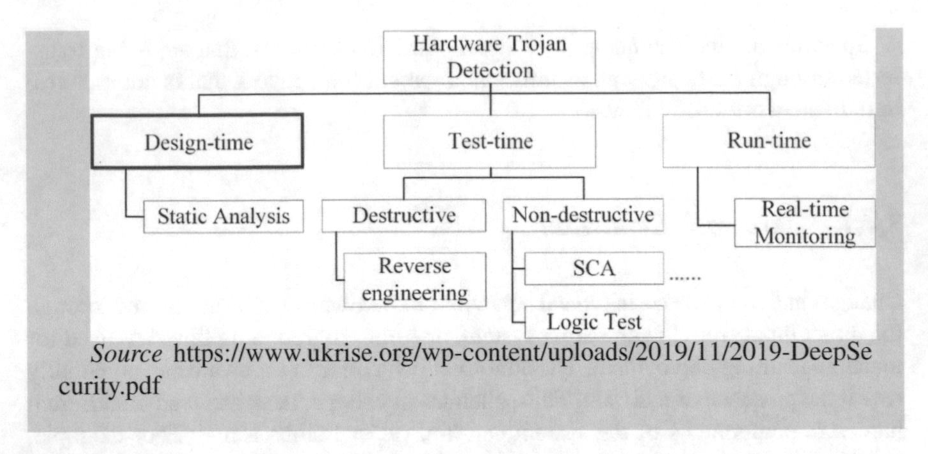

Source https://www.ukrise.org/wp-content/uploads/2019/11/2019-DeepSecurity.pdf

3.4.2 Operating System

The behaviour of the operating system is monitored through the changes in the configuration and access permissions at the level of system software.

System software is designed to provide a way for the user to interface with the device. One of the most well-known types of system software are operating systems, that allow the user to install other software and interact with additional connected devices. Examples of the system software include operating systems like macOS, GNU/Linux, Android and Microsoft Windows, computational science software, industrial automation, game engines, and software-as-a-service applications.

In contrast to the system software, software that allows users to do user-oriented tasks such as to create text documents, play games, listen to music, or browse the web are collectively referred to as application software.

Malware threatening operating systems is evolving aside all modern technologies. Among them there are banking trojans and ICS ransomware, affecting small and large businesses all over the world. A very unique type of malware, that stands out is a rootkit, which is a system-level or administrator-level malware, that hides its presence by intercepting and changing data that is being displayed to the user.

An operating system attack

Many privilege escalation exploits use vulnerabilities in the kernel or operating system processes to gain "system-level" access, known as *NT-Authority/SYSTEM* or *root*.

To protect the operation system, it is necessary to monitor changes in system files and system directories, such as *Windows* directory in Windows operating systems or */bin, /boot,* and */dev* in linux-based operating systems.

Operating system is a bridge between the installed software and connected hardware.

3.4.3 Connected Devices

Many types of devices can be directly connected to most of the computers via a variety of ports. Typically, a computer would have at least USB or USB-C ports, which can be used to bypass the security, install a keylogger, or even to wipe a computer completely. "Bad USB" attacks are easy to execute when the attacker has a physical access to the computer or can emulate the USB for a virtual container.

Ethernet ports can be used to connect "networked" devices. Connecting a malicious device would usually be categorised as the realm of network security, unless the fault is in the Ethernet port itself.

Some computers still have Serial Ports, that can be used for malicious purposes.

Power supply port can be also used as a malware injection point, however only in cases when the power supply port and (a variation of) the USB are the same. This injection is unrelated to the power supply, and directly related to the USB/USB-C attack vector.

3.4.4 Software Analysis

Analysis of the software behaviour is aimed to detect changes at the application layer of the system, while focusing on one or multiple applications, their designated directories or even virtual containers. It is worth noting, that the attempts to access abnormal processes in the system can be detected using conventional IDS. However, machine learning allows to work with sequences and sets of metadata, which makes it a perfect tool to detect repeated attempts, processes migrations, or internal scanning attempts, and even guessing if an uploaded file is an anomaly or not. Methods for code injection detection can be applied for the protection of command line against command execution and privilege escalation attempts.

There have been developed several different approaches to the application-level security testing using machine learning:

- Source code analysis
- Execution and behaviour analysis
- Fuzzing of inputs
- Reverse engineering.

Source code is analysed for the use of vulnerable or potentially vulnerable functions by comparing them to the list of known functions and publicly known vulnerabilities in them.

Execution and *behaviour* are analysed by the access attempts and data flow between the application and other services and processes in the memory and CPU of the device. With the emergence of virtual containers, the analysis of the behaviour of the applications has become more accessible and more precise.

Fuzzing of the input fields is the approach to sending malformed and junk data of different datatypes, attempting to cause any kind of malfunction or even crash of the application.

Reverse engineering is aimed to extract the original source code from the already compiler, packaged and encrypted executable file of the application. A lot of the process has to be done manually, but machine learning can greatly empower this process.

The challenge is a real-time operation of the system and the integration of the machine-learning-based IDS into that process. Every system call will make a recursive number of system calls from the IDS, in the attempts to monitor internal activity. It is not possible at a current stage of development to implement a fully functional host IDS that would process all the activity in the system, as the use of real-time machine learning will result in significant increase in required resources for the same operation. Such computational complexity prevents machine learning from being adopted as a good supplement to the already existing defence software.

Application attack

Text recognition applications can be injected though the "recognised" text, by including a line of code into the text.

3.5 AI-Specific Security Issues

Machine learning has a number of requirements to be deployed on a device with an operating system of a relatively modern version, a set of programming libraries, sufficient storage capacity if the model is large, and the IDS itself. Being a powerful tool, a system like this needs enhanced protection. With each new library or software an additional point of failure and new vulnerabilities cause the emergence of new attack vectors.

Below is the description of security issues and challenges, that can appear in using machine-learning-based intrusion detection.

3.5.1 Adversarial Attacks on Artificial Intelligence

While developing a machine-learning-based system, the training data is usually presented in the same way as the testing data, and they are identical in shape and largely similar content. In the real environment this is never the case. The samples are not balanced, the behaviour is distributed over time, and the traffic will be both malicious and non-malicious. Misclassifications of the unknown traffic will happen and it will happen very frequently. This phenomenon attracted researchers to apply *adversarial approaches* in machine learning to approximate the real-life conditions and thus expanding the boundaries of the IDS flexibility, improving performance and detection accuracy.

Adversarial approach attempts to alter the data presented to the model, intentionally aiming to make the model misclassify the data. For examples, through making the malware behaviour look more like a normal traffic; or generating a malicious payload for the code injection queries that could bypass conventional protection methods. Researchers have successfully developed systems that can introduce the pixel-level alterations invisible to a human eye, but able to cause misclassification by a model. Generating samples like this allows to create datasets with the synthetic samples being hard to classify. They are further used to train more accurate models.

In the real-life conditions, the learning process can be used to inject intentionally malicious "training" samples into the target machine-learning-based unsupervised or semi-supervised recognition or intrusion detection system. Injecting intentionally wrong statements into the training dataset reduces the detection accuracy of the system. The detection of contextually "wrong" samples is usually not possible through any currently available automated methods, and it is virtually impossible to manually detect a few corrupted statements in a flow of millions of correct statements.

The same approach can be taken to poisoning the systems that are continuously learning. One of the notorious examples is the Microsoft chatbot Tay. When the learning process gets injected with huge volumes of the noise data, the system becomes unusable very quickly. Same would happen to a self-learning intrusion detection system under a DoS flood attack.

3.5.2 Defence Methods Against Adversarial Attacks

The technology that makes adversarial attacks possible is the same technology that allows to defend against them, proving one more time the dual-use nature of information technology, and specifically of AI.

Defence against adversarial attacks consists of several key components:

- Detection of errors in input
- Detection of errors in logic
- Recovery from errors in logic
- Learning from adversarial samples.

Below are the most common examples of the use of machine learning for defence.

Anti-phishing

Phishing is a widely used method of luring users to perform an action or a series of actions, that would allow the malicious actors to access the system and then the network. With the increasing number of phishing attacks, the defence methods have remained pretty much the same—blacklisting of the links and signature-based detection of the content. Other methods, like image analysis and non-machine-learning-based natural language processing are very time-consuming and slow down the delivery of the emails.

Machine learning has proposed a number of innovative methods, and currently there are over 30 different types of machine learning models that can be applied for anti-phishing. They are specifically suitable to subject and content analysis, attachment evaluation, image recognition, the analysis of links, and so on.

Below are the attacks that are less dependent on the actions of a human operator.

Modifying training process and input data

Adversarial samples have to be introduced into the training dataset, in order for the neural network to keep improving the detection performance. Further, a live IDS can collect the unknown samples and send to the developers, or the dedicated server with the sufficient capacity to train updated models. Alternatively, if the server where IDS is deployed has sufficient computational power, the training can happen on a live system and the model can be updated with the new samples almost in real time.

Modifying models

The combination of adversarial samples and constantly evolving internal architecture of the machine learning models allows to improve the performance, learning rate, and robustness of the intrusion detection systems.

In the vast majority of cases, when the model needs to have configurations changed or modified, a completely blank model would have to be trained instead. For example, if the number of neurons in a deep neural network is changed, then a blank model would have to be declared, trained, and evaluated.

Using an additional model

Every model has limits in accuracy and predictive capabilities. A separate training model alongside the main model may be required, as it can generate new samples from the latest version of the main model and keep improving it even further.

Some of the additional models can be hosted on remote systems. This approach can reduce the computational requirements for the intrusion detection, as the model that is trained can be placed to a separate high-performance computer or a cloud, and it can be evaluated for defects and errors, before replacing the currently active model.

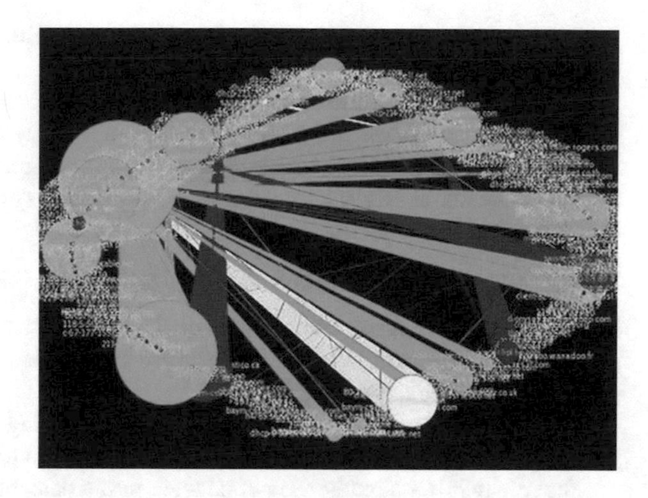

Fig. 3.2 Traffic flow between computers in the network during (Screenshot of EtherApe, quality is intentionally lowered)

Cyber-attack visualisation

The visualisation of the model output results is vital for the early detection of upcoming and ongoing cyber attacks. With thousands of alerts, generated by IDS, the potential cyber attacks have to be also mapped and laid out on a network topology. It is especially crucial for a network that spans across a country, a continent, or even the world (Fig. 3.2).

In large-scale networks with high number of nodes and overwhelming amount of false alarms it is extremely difficult, and in most cases, virtually impossible, to detect sophisticated attacks. New approaches are needed that will allow for more computational power and training options to improve the visual representation of the attacks, to automatically detect false positive alarms, and present only relevant events to operators.

3.5.3 Development of Safe Artificial Intelligence Systems

Even though machine learning has not yet been widely adopted, certain best practices aimed at developing safe AI systems have already been applied in the developer community.

By default, best practices for secure hardware and software development apply first. There is a variety of standards and recommendations for secure development, at the national and international levels, including industry-specific frameworks.

Safe distributed machine-learning-based software

Research is conducted in the development of the distributed machine learning in a cloud where multiple participants can contribute their datasets and computational power to model training, without sharing the datasets with each other, but sharing a model trained with combined efforts.

Such method can contribute to the development of potentially viable solutions to privacy preservation, since the content of the trained model is irreversible. But the model would be highly vulnerable to dataset poisoning attacks and insider threat. For such method to work, a model or a quality-check system has to be developed that can detect false training samples.

Machine learning classification over encrypted data

The majority of modern communication in the Internet is encrypted. Industrial communication protocols have also received an upgrade to provide integrity and confidentiality of transferred data.

Some researchers have attempted to train models using encrypted data. Machine learning can detect those correlations, that seem random to human eyes.

The research has already been conducted in the detection of malware and cyber attacks over encrypted channels, however this type of detection requires highly informative features about the network flow and complex machine learning models to be successful. These methods are typically applied for encrypted traffic classification, but they have also been applied to crypto-mining malware detection.

3.5.4 Hybrid Defence

Signature-based and anomaly-based intrusion detection are significantly different in effectiveness, accuracy, and levels of integration into the networks and systems. In order to maximise the effectiveness of security measures, both methods have to be used at the same time.

The location and the order of execution are based on the threat model for a given network or group of networks. For example, malware for ICS can be unique and the signatures—not present in the anti-virus software and firewalls. Thus, in industrial networks anomaly detection might be more valuable, than a detection of specific signature.

A continuously increasing number of networked sensors, monitoring and support software and devices, are blurring the border between virtual and physical worlds. In the near future the number of sensors will be growing, and the integration between two worlds will keep expanding.

3.6 Conclusion

The overview of the defence methods provided in this Chapter has demonstrated that machine learning has a promising potential in cyber defence, and multiple advantages over conventional methods. The major areas of its current application are anomaly and misuse detection.

Machine learning application in cyber security is still at an early stage and has not reached levels sufficient to be considered mainstream. Among the reasons are its high costs, need in large data and high skill experts, lack of understanding of its decision-making processes, multiple errors and false positives, lack of autonomy and privacy, specific vulnerabilities.

At the design stage, the developers have to follow standards, recommendations, and best practices for secure software development. However, the standards for machine learning have not been fully developed and approved yet due to complexity of the technological process and its variability.

In defence missions, machine learning is aimed at helping humans to protect computers and networks, and promises to replace human operators, especially in round the clock activities. Further research of its imperfections and limitations will undoubtfully result in their elimination, contributing to a broader use of this technology.

To be more efficient in defence, the knowledge of the attacks characteristics and their potential is essential. The AI-enhanced cyber attacks, their specificity in relation to machine learning and growing threats will be discussed in the next Chapter.

Chapter 4
Attack

Cyberspace was created to enhance information exchange and eliminate communication distances. However, from the moment when the cyberspace was created, it has been also used for malicious operations and adversarial attacks. Currently, it is considered as one of the military domains, together with the land, air, sea and space. Such terms as cyber operations, cyber conflicts and even cyber wars are now widely in use.[1]

> Recent years have seen significant advances in a wide array of new and emerging technologies with disruptive potential, several of which have an **inherent cyber dimension**. These include, inter alia, artificial intelligence and machine learning, autonomous devices and systems, telecommunications and computing technologies, satellites and space assets, human–machine interfaces and quantum computing.[2]
>
> *Cyber Threats and NATO 2030: Horizon Scanning and Analysis, 2020*

AI brought advantages not only to science and peaceful application, it also empowered malicious actors, using AI algorithms to negatively influence the performance and disrupt the normal functions of non-malicious programmes. Among the characteristics of the AI-empowered cyber attacks, there are higher levels of evasiveness,

---

[1] Bellasio, J., Silfversten, E., Leverett, E., Quimbre, F., Knack, A. & Favaro, M. (2020) The future of cybercrime in light of technology developments. Prepared for the European Commission Structural Reform Support Service (Ref: SRSS/ C2018/092).

[2] Jacopo Bellasio, Erik Silfversten, "The Impact of New and Emerging Technologies on the Cyber Threat Landscape and Their Implications for NATO", in Cyber Threats and NATO 2030: Horizon Scanning and Analysis', ed. by A. Ertan, K. Floyd, P. Pernik, Tim Stevens, CCDCOE, NATO Cooperative Cyber Defence Centre of Excellence, King's College London, Williams and Mary, 2020, p. 88 at https://ccdcoe.org/uploads/2020/12/Cyber-Threats-and-NATO-2030_Horizon-Scanning-and-Analysis.pdf.

S. Abaimov and M. Martellini, *Machine Learning for Cyber Agents*,
Advanced Sciences and Technologies for Security Applications,
https://doi.org/10.1007/978-3-030-91585-8_4

scaling potential, customisation and personalisation. Cyber attackers apply the benefits of machine learning to reinforce **cyber arms**, such as botnets, spearfishing, and evasive malware. Deep fakes make viewers believe in the video and audio fabricated information, spread misinformation and disinformation, manipulate society opinion, interrupt conferences, political debates, distort interpretation.

AI greatly improves the attackers' capabilities in passive and active reconnaissance, generation of exploitation payloads, traffic masquerading, creating phishing emails, or causing physical damage to systems. Cyber attacks are of a particular danger to the autonomous systems when their operational process may be manipulated. The number of AI-powered attack tools will only grow in the future, as machine learning is gaining popularity and accessibility.

Expected economic losses from cyber attacks

The bill for the many depredations of the 2017 NotPetya attack was roughly $10 billion. [...] To economists who routinely put a monetary cost on life in making cost–benefit calculations, many cyber attacks are more serious than military confrontations short of fully committed war.[3]

Cyber Threats and NATO 2030: Horizon Scanning and Analysis

If it were measured as a country, then cybercrime—which is predicted to inflict damages totaling $6 trillion USD globally in 2021—would be the world's third-largest economy after the U.S. and China. Cybersecurity Ventures expects global cybercrime costs to grow by 15% per year over the next five years, reaching $10.5 trillion USD annually by 2025, up from $3 trillion USD in 2015.[4]

2021 Report: Cyberwarfare in the C-suite

The most dangerous scenario in the perceivable future is the AI weaponisation when the technological advancements empower weapons systems allowing their autonomy and thus independence, including in taking critical "life or death" decisions. They are used on the land, in the air, water and space, with the nuclear and chemical weapons being of existential danger.

[3] 'Cyber Threats and NATO 2030: Horizon Scanning and Analysis', ed. by A. Ertan, K. Floyd, P. Pernik, Tim Stevens, CCDCOE, NATO Cooperative Cyber Defence Centre of Excellence, King's College London, Williams and Mary, 2020, p. 67 at https://ccdcoe.org/uploads/2020/12/Cyber-Threats-and-NATO-2030_Horizon-Scanning-and-Analysis.pdf.

[4] Steve Morgan, Cyber Security Ventures, *2021 Report: Cyberwarfare in the C-suite, 2021, p. 1, at* https://1c7fab3im83f5gqiow2qqs2k-wpengine.netdna-ssl.com/wp-content/uploads/2021/01/Cyberwarfare-2021-Report.pdf.

> Then there could be an algorithm that said, "Go penetrate the nuclear codes and figure out how to launch some missiles." If that's its only job, if it's self-teaching and it's just a really effective algorithm, then you've got problems. I think my directive to my national security team is, don't worry as much yet about machines taking over the world. Worry about the capacity of either nonstate actors or hostile actors to penetrate systems, and in that sense, it is not conceptually different than a lot of the cybersecurity work we're doing.[5]
>
> *Barack Obama, 2016*

There are also potential scenarios of weaponising the civil AI-empowered devices targeting individuals or organizations, or even states. A malware that targets widely used services and hardware, such as IDS or IoT devices, would destabilise regions or even countries. The well-known example of the first ever machine-learning-based malware is Deep locker, ransomware developed in 2018 by the IBM Research lab. It demonstrates highly evasive capacities and remains hidden until it reaches the target.

The conducted literature review has demonstrated scarcity in the analysis of attack methods with the use of machine learning. Thus, as of today there are only a dozen of them that were developed for research purposes and that demonstrate the AI attacking capacity (see Sect. 4.2.2).

This chapter aims at contributing to the body of knowledge and filling in the existing gaps in researching the use of machine learning for malicious purposes. It will provide an overview of the machine learning application in enhancing cyber arms and their use in sophisticated attacks.

4.1 Machine Learning for Malware

Cyber attacks are implemented by **cyber arms**, programs created with the malicious purposes, e.g. to steal information, interrupt business processes, cause damage, etc. A cyber arm is "a software able to function as an independent agent and run commands. It has a dual use nature and can be used for both offence and defence purposes."[6]

Malicious software, or **malware**, is a specific cyber arm used for malicious purposes. It is specifically designed to disrupt, damage or gain unauthorized access to a computer system or network. Depending on the purposes and proliferation systems, malware can be divided into various, not mutually exclusive categories. The historical classification assigned names to specific malware based on its functions and capacities. The long list, that is far from being exclusive, enumerates viruses, worms,

[5] Barack Obama, Neural Nets, Self-driving cars, and the future of the world, 24 August 2016, https://www.wired.com/2016/10/president-obama-mit-joi-ito-interview/.

[6] Abaimov and Martellini [1].

trojans, ransomware, rootkits, downloaders, backdoors, etc.[7] Modern malware is more "all-in-one" and may combine capabilities of several groups.

Malware history has started from the creation of the first malicious programmes back in the 1970s, prank viruses that were mostly entertaining the users and researchers, while not causing any real impact. Since that time, they have evolved and are now able to self-replicate, evade detection and cause a targeted and wide-spread impact.[8]

From the history of evasive malware

- **Late 80s-early 90s**: polymorphic and metamorphic malware appears. It was able to avoid signature analysis and thus pushed the development of antivirus software further, to analyse code and detect the intent of malware.
- **1990s**: malware developers started using encryption for malware binary executable files. In response the defenders stated using sandboxing technology to quarantine suspicious files and analyse there behavior.
- **2000s**: malware attempts to actively avoid analysis, for example, refusing to execute the payload in virtualised environment. This approach is still relevant and widely used.
- **2010s**: the trend has changed to targeted malware, that activates the payload only when hosted on specific hardware or when specific software is present (e.g., SCADA).

Malware in the AI era will be able to mutate into thousands of different forms once it is lodged on a computer system. Such mutating polymorphic malware already accounts for more than 90% of malicious executable files. Deep RL tools can already find vulnerabilities, conceal malware, and attack selectively. While it is uncertain which methods will dominate, there is a clear path for U.S. adversaries to transform the effectiveness of cyber attack and espionage campaigns with an ensemble of new and old algorithmic means to automate, optimize, and inform attacks. This goes beyond AI-enhanced malware. Machine learning has current and potential applications across all the phases of cyber attack campaigns and will change the nature of cyber warfare and cyber crime. The expanding application of existing AI cyber capabilities will make cyber attacks more precise and tailored, further accelerate and automate

[7] For a more detailed list of malware and its capacities, see Abaimov S., Martellini M. Cyber Arms. Security in cyberspace, CRC Press, 2020.

[8] *For more information see* Abaimov S., Martellini M. Cyber Arms. Security in Cyberspace, CRC Press, 2020.

cyber warfare, enable stealthier and more persistent cyberweapons, and make cyber campaigns more effective on a larger scale.[9]
Final Report—National Security Commission on Artificial Intelligence, 2021

Machine-learning-powered malware has no comprehensive reasoning and is efficient at what it is designed to do. There is a potential that an intelligent malware would keep evolving and improving itself into an advanced cyber weapon that cannot be stopped.

I am deeply concerned about advanced forms of malware. We're not there today yet. But you could envision things over time that are adapting and learning and begin to populate the web, like there are people doing interesting ways of thinking about systems that have misaligned goals. It's also possible to envision systems that don't have any human directed goals at all. Viruses don't. They replicate. They're effective at replicating, but they don't necessarily have a goal in the way that we think of it other than self-replication.[10]

Paul Scharre, 2020

Many modern malware are file-less, and have no physical presence on the storage device of the computer system. The machine-learning-powered malware is hypothetically able to avoid the analysis of the executable files and is able to propagate without triggering the intrusion detection system. The malware that learns by itself requires innovative approaches to intrusion detection, i.e. the combination of signature-based and anomaly detection.

Machine learning can be used to generate either malware source code or a full binary executable. Code payload generation usually needs a list of possible malicious and non-malicious payloads, and a way to test them for syntax accuracy. To look indistinguishable from non-malicious files, the malicious process can imitate any system process or an application level subprocess, that has the non-malicious behavioural properties, including the file details and credentials, as well as the lower-lever behavioural patterns. That requires a malware developer to have a sample of the process file, and ideally a source code of the executable component of the process, for example, *svchost.exe* in Windows or *syslogd* in Linux.

Antivirus software detects malware by matching the code to a known "signature", which is a code sample or a pattern. As an executable file of a conventional malware is not yet capable of changing itself to avoid detection by signature-based anti-virus

[9] Final Report—National Security Commission on Artificial Intelligence, March 2021, p. 50, United States of America, https://www.nscai.gov/wp-content/uploads/2021/03/Full-Report-Digital-1.pdf.
[10] Paul Scharre, Interview to the Future of Life Institute, March 16, 2020, (by Lucas Perry) https://futureoflife.org/2020/03/16/on-lethal-autonomous-weapons-with-paul-scharre/#:~:text=Paul%20Scharre%3A%20Yes%2C%20so%20autonomy,for%20some%20period%20of%20time.

suites, it is always encrypted and packaged. However, modern techniques allow to write or even rewrite the malware code on the flow to make it undetectable by the automatic code analysis, even if the malware executable is unencrypted and not packaged.

Malicious actors use direct and indirect functionality of any technology. For example, hiding the encryption or decryption key in the model is a secondary function, that is not directly related to the full spectrum of the learning capacities of the machine learning, yet it serves as a tool in a successful ransomware attack.

First Machine Learning malware: DeepLocker and EvilModel evasive capacities

IBM Research developed DeepLocker to better understand how several existing AI models can be combined with current malware techniques to create a particularly challenging new breed of malware. This class of AI-powered evasive malware conceals its intent until it reaches a specific victim. It unleashes its malicious action as soon as the AI model identifies the target through indicators like facial recognition, geolocation and voice recognition. [...]

DeepLocker hides its malicious payload in benign carrier applications, such as a video conference software, to avoid detection by most antivirus and malware scanners.

What is unique about DeepLocker is that the use of AI makes the "trigger conditions" to unlock the attack almost impossible to reverse engineer. The malicious payload will only be unlocked if the intended target is reached. It achieves this by using a deep neural network (DNN) AI model.

The AI model is trained to behave normally unless it is presented with a specific input: the trigger conditions identifying specific victims. The neural network produces the "key" needed to unlock the attack. DeepLocker can leverage several attributes to identify its target, including visual, audio, geolocation and system-level features. As it is virtually impossible to exhaustively enumerate all possible trigger conditions for the AI model, this method would make it extremely challenging for malware analysts to reverse engineer the neural network and recover the mission-critical secrets, including the attack payload and the specifics of the target. When attackers attempt to infiltrate a target with malware, a stealthy, targeted attack needs to conceal two main components: the trigger condition(s) and the attack payload.

DeepLocker is able to leverage the "black-box" nature of the DNN AI model to conceal the trigger condition. A simple "if this, then that" trigger condition is transformed into a deep convolutional network of the AI model that is very hard to decipher. In addition to that, it is able to convert the concealed trigger condition itself into a "password" or "key" that is required to unlock the attack payload.

Technically, this method allows three layers of attack concealment. That is, given a DeepLocker AI model alone, it is extremely difficult for malware analysts to figure out what class of target it is looking for. Is it after people's faces or some other visual clues? What specific instance of the target class is the valid trigger condition? And what is the ultimate goal of the attack payload?[11] (Fig. 3.1)

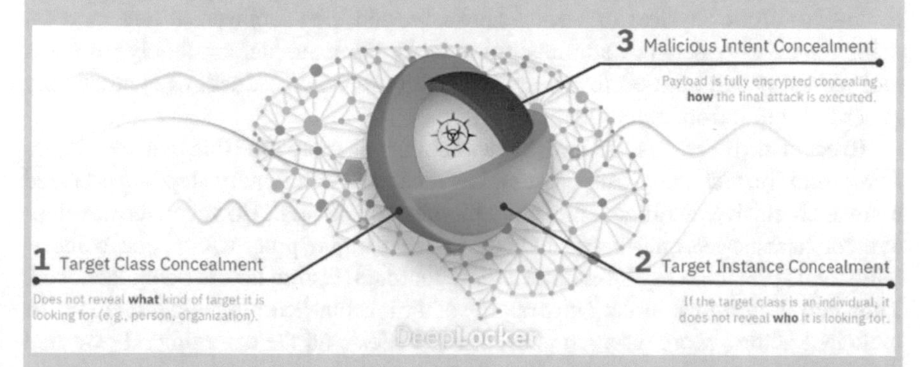

Fig. 4.1 DeepLocker—AI-powered concealment

Stoecklin, Jang, Kirat, DeepLocker: How AI Can Power a Stealthy New Breed of Malware, 2018

Experiments show that 36.9 MB of malware can be embedded in a 178 MB-AlexNet model within 1% accuracy loss, and no suspicion is raised by anti-virus engines in VirusTotal, which verifies the feasibility of this method. With the widespread application of artificial intelligence, utilizing neural networks for attacks becomes a forwarding trend.[12]

Wang, Liu and Cui, EvilModel: Hiding Malware Inside of Neural Network Models, 2021

There exist a number of complicating factors in developing machine-learning-based malware. For example, malware testing requires a specific environment and goes beyond just checking it for detectability, as one would do with a non-machine-learning malware. Malware can create unpredictable and damaging effect, as it should, and subsequently damage the hosting system itself.

Literature review shows a number of innovative machine learning methods used by researches to generate malware samples, in order to better understand and develop the defence mechanisms. To address the challenges of testing and verification, genuine and synthetic samples of malware can be deployed in virtual containers to avoid

[11] Marc Ph. Stoecklin, Jiyong Jang, Dhilung Kirat, DeepLocker: How AI Can Power a Stealthy New Breed of Malware, 8 August 2018, at https://securityintelligence.com/deeplocker-how-ai-can-power-a-stealthy-new-breed-of-malware/.

[12] *Zhi Wang, Chaoge Liu, Xiang Cui, EvilModel: Hiding Malware Inside of Neural Network Models, July 2021, at.* https://arxiv.org/abs/2107.08590.

contamination of the host operating system. The behaviour of the malware can be observed using monitoring tools and anti-virus software, and tested for detectability using various unconventional intrusion detection methods.

As an alternative to malware execution in physical systems, the real-life scenarios can be simulated or replayed from traffic records. The network traffic can be recorded in advance from physical or virtual networks, and then replayed at any moment using the existing network administration tools. For example, a publicly available tool *tcpreplay* can be used to replay the already recorded, edited, or synthetically generated traffic from *pcap* files.

To conclude the review, it is worth mentioning one more time that malware, benefiting from the technological progress, has made an evolutionary step. AI-powered malware is elusive, intelligent and has a high threat potential. However, the development of machine-learning-based malware is a challenging process, e.g., due to uncertainty in its predictability it requires a more thorough testing before being launched. Thus, currently, only indirect functionality of the machine learning is exploited by the malicious actors. More research is needed to understand the emerging AI-powered cyber attacks and to be subsequently able to prevent them.

4.2 Machine Learning Enhancing Cyber Attacks

The AI-enabled tools allowed to launch innovative and enhanced attacks against conventional and machine-learning-based systems. Literature review of the publicly available sources has demonstrated that the research in the use of machine learning for cyber attacks requires more attention. The scarcity of this research may be partially explained by the fact that at the current stage there are no publicly known reported cases of machine learning being used to directly attack a system or an application. But it can be used in a variety of ways that can increase the capabilities of the attackers. This includes the generation of executable files of malware, payloads of code, and generation of synthetic purpose-specific traffic records and behaviour.

> From an adversarial perspective, AI/ML could be leveraged for nefarious purposes to automate cyber attacks. While the use of AI/ML for such purposes has not yet been observed in the wild, companies have already launched 'red teaming as service' platforms offering automated attack services which

combine a confidence engine with target temptation analysis to detect system and network vulnerabilities, highlighting assets with the highest perceived adversarial value.[13]

Cyber Threats and NATO 2030: Horizon Scanning and Analysis

Machine learning as a technology enables attackers with capacities that would have otherwise been impossible. It has enhanced the already existing conventional attacks, such as active reconnaissance, exploitation and botnet management, pattern recognition in cryptanalysis, and forensics evasion. Improving the performance of network scanners and vulnerability assessment frameworks (OpenVAS, Metasploit, etc.) would make them more efficient, allow collecting more accurate information about the target systems with minimal footprint.

Machine learning models can be used as a primary or a secondary tool in cyber attacks. As a **primary tool**, they are directly used for exploitation, while as a **secondary tool**, the model is used to **enable**, **improve**, or **supplement** the actions of the attacker or malware. As a machine learning model can be used as a primary or as a secondary attack tool, the combinations for attacks are limitless. In addition, the use of machine learning models for defence enables attackers to specifically target their vulnerabilities, e.g. through poisoning data.

Activities that do not fit into two previous categories are additional approaches, such as indirect optimisation of performance by reducing the size of password lists, load balancing of DDoS tools, use of social media scraping to improve the quality of phishing emails. Secondary tasks and activities can improve the efficiency of the attack during preparation stages and during steps that require offline or indirect data processing, that does not involve the target systems and networks.

Malicious actors can use machine learning to mimic their behaviour or masquerade the behaviour of the malware as a non-malicious software or traffic. Behaviour mimicking has already been extensively studied and implemented in numerous proof-of-concept scripts and programs. Traffic behaviour impersonation is not directly an attack, but a way to hide an attack against anomaly-based intrusion detection methods. Machine learning model learns from the traffic flow, and creates an approximation of how the malicious packets should be sent to the target system or network, in order to avoid being detected. Attackers can design a more advanced system able to create network tunnels, that would forward any malicious traffic through those tunnels with frequency and behaviour of a non-malicious everyday activity in the target network.

[13] Jacopo Bellasio, Erik Silfversten, "The Impact of New and Emerging Technologies on the Cyber Threat Landscape and Their Implications for NATO", in Cyber Threats and NATO 2030: Horizon Scanning and Analysis', ed. by A. Ertan, K. Floyd, P. Pernik, Tim Stevens, CCDCOE, NATO Cooperative Cyber Defence Centre of Excellence, King's College London, Williams and Mary, 2020, p. 90 at https://ccdcoe.org/uploads/2020/12/Cyber-Threats-and-NATO-2030_Horizon-Scanning-and-Analysis.pdf.

Evolving cyber offensive capacities can lead to the existing defence techniques being rendered incapable to prevent a constantly growing volume of malicious traffic from compromising the integrity of any organisation. The increasing complexity of attack tools challenges the defence that has to protect the infrastructure without affecting the accessibility of the services for users and consumers.

Cyber attack is a "one-sided offensive activity in the computer system or network to violate confidentiality, integrity, and availability of information, disrupt services, illegally use or destroy the system."[14] In general, a conventional cyber attack as a phenomenon consists of the following stages: reconnaissance, scanning and vulnerability assessment, breach and exploitation, assault, exfiltration, persistence, obfuscation, covering tracks.[15] Each of those stages can be enhanced with the AI in one way or another. For example, reconnaissance can be enhanced by collecting more information about the selected target network or organisation, scanning can be improved by optimising the footprint and making it almost indistinguishable from a legitimate query.

> There is reason to expect attacks enabled by the growing use of AI to be especially effective, finely targeted, difficult to attribute, and likely to exploit vulnerabilities in AI systems.[16]
>
> *The Malicious Use of Artificial Intelligence, Report, 2018*

The attack types vary and their volume is skyrocketing. Machine learning can be used to enhance phishing, password guessing, ransomware, and almost all other types of cyber attacks. While conventional attacks are guided by very strict logical algorithms, the AI-enhanced attacks are more flexible. Conventional attacks are more precise while the AI attacks are more versatile and adaptive to the target requirements.

Cyber attacks can be categorized by steps in the attack methodology, by impact, type of malware, the level of covertness, by the level of automation, by vulnerabilities exploited. Most commonly they are classified by the impact and technique used. For the purpose of this book, cyber attacks are categorised by the attack method. Table 4.1 illustrates how machine learning can enhance those conventional attacks.[17]

Machine learning models can be used as attack vector, as a standalone attack tool, or can enhance conventional attack methods. In this subchapter we present those conventional attacks in a more detailed way.

---

[14] *Abaimov S., Martellini M. Cyber Arms, 2020.*

[15] For more detail, see Abaimov S., Martellini M., 2020, pp. 18–26.

[16] Miles Brundage et al., Report on The Malicious use of artificial Intelligence: Forecasting, Prevention and Mitigation, February 2018, last retrieved 15 July 2018 at http://img1.wsimg.com/blobby/go/3d82daa4-97fe-4096-9c6b-376b92c619de/downloads/1c6q2kc4v_50335.pdf.

[17] *Of note: only the major types of attacks are presented.*

Table 4.1 Enhancement of cyber attacks using machine learning

Attack	Machine learning enhancement is used to:
Phishing	Maximise the total number of clicks
Spear-phishing	Personalise the email to maximise the chance of the link activation by one target user
Exploitation	Generate a functional code-injection query that can avoid filters and detection
Network traffic masquerading	Impersonate genuine non-malicious traffic
Bots and Botnets	Optimise propagation and control of bots
Password guessing	Password dictionary generation
Ransomware	Conceal decryption key in the model
Cryptomining	Enhanced stealth and more efficient mining
Recovery	Deleted data detection and recovery
Cryptanalysis	Pattern detection and faster decryption
Forensic investigation	Enhanced stealth against forensic software
Hardware attacks	Pattern recognition for reverse engineering

4.2.1 Phishing

Phishing is the most common type of cyber attacks, that exploits the human component in the system and is a lot harder to defend against, as compared to other vulnerabilities.

Phishing is an attack, and requires human decision making to select suitable targets for spear-phishing or compose trustworthy emails for mass distribution. Machine learning has already been applied for the candidate selection in job interviews. A very similar tool can be used to automatically select an ideal target for a spear-phishing attack. Similarly, natural language processing can be used to generate more trustworthy phishing and spear-phishing content.

Automated Customized Social Engineering Attacks

Many major cybersecurity failures began with "social engineering," wherein the attacker manipulates a user into compromising their own security. Email phishing to trick users into revealing their passwords is a well-known example. The most effective phishing attacks are human-customized to target the specific victim (aka spear-phishing attacks)—for instance, by impersonating their co-workers, family members, or specific online services that they use. AI technology offers the potential to automate this target customization, matching targeting data to the phishing message and thereby increasing the effectiveness of social engineering attacks. Moreover, AI systems with the ability to create realistic, low-cost audio and video forgeries will expand the phishing attack space from email to other communication domains, such as phone calls and video conferencing.[18]

Michael C. Horowitz et al. (incl. Scharre\), Artificial Intelligence and International Security, 2018

In phishing the machine learning model is trained prior to the attack itself, improving the quality or automating the creation of malicious emails or advertisements. Dataset can be composed using genuine emails that contain requests to action, such as opening an attachment, activating a link, or downloading a file from a cloud. After analysing a set of legitimate emails and comparing them to intentionally malicious phishing emails, a generative machine learning model can create new text samples, that attempt to bypass a spam filter or a phishing detection system.

The challenge of this approach is to generate a comprehensive message. A perfectly crafted phishing email that successfully bypasses spam filters, can in reality be a pile of words, that would make no sense to a person who receives the phishing email. Thus, an additional context-aware filter is required to verify the quality of the generated text samples.

Advertisement content would be a lot simpler to produce using machine learning and to further automate the entire "supply chain". A fully automated phishing tool without the need to create content of the emails or advertisements could cause a "cyber pandemic".

All the machine learning models, that are suitable for the natural language processing, are equally suitable for phishing email generation.

Spear fishing, enhanced with AI, is able to collect personal data in big scales. Researchers have already demonstrated that a spear phishing kit could create custom messages, using Twitter preferences as an input data, attracting a high number of clicks, that could have led to an injected website.

Existing frameworks (e.g., Social Engineer Toolkit) can be used to automate the generation of the payload for phishing. However, those frameworks are very limited in tailoring the content to a specific target.

When a phishing attack is not possible, the attackers have to rely on the vulnerabilities in the software used by the target organisation.

[18] Michael C. Horowitz et al. (incl. Scharre), Artificial Intelligence and International Security, July 2018, Center for a New American Security's series on Artificial Intelligence and International Security, page 3 https://ethz.ch/content/dam/ethz/special-interest/gess/cis/center-for-securities-studies/resources/docs/CNAS_AI%20and%20International%20Security.pdf.

4.2.2 Exploitation

Machine learning can enhance the identification of vulnerabilities and their exploitation techniques. Furthermore, reconnaissance, vulnerability detection, and exploit development can be designed in such a way, that a machine-learning-based system can detect zero-day vulnerabilities and develop own exploits or at least assist the malicious actors in developing zero-day exploits.

Exploit

A constructed command or a software designed to take advantage of a flaw in a computer system (vulnerability), typically for malicious purposes, such as accessing system information, establishing remote command line interface, causing denial of service, etc.

Abaimov, Martellini, 2020, p. X

The new technology contributes to the enhancement of the attack methods though optimisation, new approaches to previous attacks, and innovative attack patterns. Fuzzing techniques and code injection techniques are those that can benefit from the machine learning the most.

However, machine learning as a new technology has its own multiple vulnerabilities that can be exploited.

4.2.2.1 Exploiting Machine Learning Vulnerabilities

Virtually all the types of machine learning are susceptible to some form of attacks, which ultimately lead to misclassification in detection by the model. As mentioned in Sect. 2.3, machine learning can be divided into data collection, pre-processing, training, evaluation, prediction, and visualisation. In each of those stages the data or the model can be poisoned, corrupted, or tampered with. Data collection and data generation stage is specifically susceptible to data poisoning, small alterations in pre-processing change the learning accuracy completely, and undertraining or overtraining cause higher rate of false predictions.

Table 4.2 shows selected approaches to corrupting functionality of the non-malicious AI-based software.

In the presented list all the created interferences and attacks had a high success rate and managed to cause misclassification by the models. Those selected algorithms show a variety of ways through which the model can be misguided and potentially exploited.

In defence applications, in the adversarial methods, the malicious samples are becoming continuously harder to detect. Even though this method is used to train

Table 4.2 Malicious AI algorithms used to tamper data for bypassing defencive AI algorithm source[19]

	Year	Target	Impact
1	2017	Traffic signs	Misclassification of the traffic sign by AI algorithms, which can lead to traffic accidents in autonomous cars
2	2018	Medical image data	Misclassification of medical abnormalities by AI algorithms which can lead to false diagnostics of health conditions
3	2018	Facial image data	Misclassification of face images, which can lead to authentication bypass in certain scenarios
4	2019	Digital recommendation system	Data poisoning to AI algorithms, which results in wrong recommendations
5	2019	Computer Tomography Scan Data	Misclassification of tampered computer tomography scan 3D images, which can lead to false diagnostics
6	2019	Speech audio data	Adversarial attack on voice activated personal assistance, which can tamper their functionality
7	2020	Network Intrusion detection system	Adversarial traffic generation to bypass the security of AI-powered network intrusion detection system

better intrusion detection models, the generated harder-to-detect samples can be used for cyber attacks.

Table 4.3 presents known enhanced by machine learning tools, that have already been developed and can be used for offensive purposes. Those methods can be used for further development of a new generation of attack tools and techniques. With automation, multiple machine learning techniques can be used for the development of attack tools and malware. One of the automated ways of vulnerability discovery is fuzzing.

4.2.2.2 Fuzzing

Fuzzing is an automated technique of forwarding of malformed data to the input of a program or an application in order to cause a malfunction. It can be used by the developers to improve the security of a program and by the malicious actors to detect vulnerabilities.

[19] Yamin, M. M., Ullah, M., Ullah, H., & Katt, B. (2021). Weaponized AI for cyber attacks. Journal of Information Security and Applications, 57, 102,722. https://doi.org/10.1016/j.jisa.2020.102722.

Table 4.3 AI-powered tools that use data analysis for offensive cyber operations.[20] Updated by the authors

	Year	Name	Usage
1	2017	DeepHack	AI-powered tool to generate injection attack patterns for database applications
2	2018	DeepLocker	AI-powered tool that emulates an APT for launching complex cyber attacks
3	2018	GyoiThon	AI-powered tool for information gathering and automatic exploitation
4	2018	EagleEye	AI-powered tool for social media information reconnaissance using facial recognition algorithms
5	2018	Malware-GAN	AI-powered tool used for generation of malware that can bypass security detection mechanisms
6	2019	uriDeep	AI-powered tool that generates fake domains for usage in different attack scenarios
7	2019	Deep Exploit	AI-powered tool that automates Metasploit for information gathering, scanning, exploitation and post exploitation
8	2019	DeepGenerator	AI-powered tool to generate injection attack patterns for web applications
9	2021	*Unnamed*[21]	Mimicking consumer behaviour using machine learning
10	2021	*Unnamed*[22]	Model replication attack in IoT in energy sector
11	2021	EvilModel[23]	Model is used to hide malware

Fuzzing has four main stages, and all of them can be enhanced using machine learning:

- testcase generation
- program execution
- runtime status monitoring
- analysis of crashes.

The *testcase generation* is focused on the creation of seed files or inputs. The seed input is the original sample, that further mutates in various ways to create testcases. To validate or discard the generated testcases, a testcase filter is used. Testcase filters used various mathematical fitness functions. The testcases can be mutation-based and

[20] *ibid.*

[21] L. Cui et al., "A Covert Electricity-Theft Cyber-Attack against Machine Learning-Based Detection Models," in IEEE Transactions on Industrial Informatics, https://doi.org/10.1109/TII.2021.3089976.

[22] Qiong Li, Liqiang Zhang, Rui Zhou, Yaowen Xia, Wenfeng Gao, Yonghang Tai, "Machine Learning-Based Stealing Attack of the Temperature Monitoring System for the Energy Internet of Things", Security and Communication Networks, vol. 2021, Article ID 6,661,954, 8 pages, 2021. https://doi.org/10.1155/2021/6661954.

[23] Wang, Zhi, Chaoge Liu, and Xiang Cui, "EvilModel: Hiding Malware Inside of Neural Network Models." arXiv preprint arXiv: 2107.08590 (2021).

generation-based. Mutation-based testcases are the modified versions of the original testcases, while generation-based testcases are based on the predefined format or template and does not contain any mutation functions.

During the *program execution* stage, the instance of a program is launched with mutated or generated inputs. As it takes time to load a program into memory, send an input, and monitor, multiple instances of the same executable file are launched to increase the overall speed of fuzzing.

Each launched instance of the program has to be analysed at runtime, as an input might cause a crash (emergency stopped execution) or the program would proceed unaffected. If the program allows execution of the input data sample, the sample should be discarded and the program instance should be terminated to free space for another cycle of instances. If the input causes a "crash" (e.g., the program halts), the crash-report is saved along with the executable file and the input data sample.

Each crash is visually similar, as it results in the program halt, when the CPU stops processing the operations from the program. Collected crashes (saved as "file states") have to be analysed, as in the majority of cases at the binary level the code execution malfunction is different in every case.

The final goal of fuzzing is to redetect as many crashes as possible and then to fix every case, improving the security of the program. Attackers can use the same set of tools to detect buffer overflow and code injection points.

Publicly available datasets for fuzzing that can be used for machine learning research include:

- "Big Code" dataset
- NIST SARD project dataset
- GCC test suite dataset
- DARPA Cyber Grand Challenge dataset
- LAVA-M dataset
- VDiscovery dataset[24]

Wang et al., 2019

Machine learning can be applied at different stages of fuzzing, namely at:

- seed file generation
- testcase generation
- testcase filter
- mutation operator selection
- fitness function
- exploitability analysis of each crash.

[24] Wang, Y., Jia, P., Liu, L., & Liu, J. (2019). A systematic review of fuzzing based on machine learning techniques. PLoS ONE, 15 (8 August). https://doi.org/10.1371/journal.pone.0237749.

Machine-learning-based fuzzing tools cannot yet compete with conventional fuzzing tools. However, they have already shown promising potential to outperform conventional approaches. The introduction of machine learning improves coverage, unique code path, unique crashes, performance and efficiency of fuzzing.

There is a very limited number of publicly available datasets that can be used for fuzzing, and those that exist might contain less features, samples, categories, and more biases. Custom-made datasets can be composed from the data collected by web crawlers and from already existing fuzzers, however, those datasets are not universally applicable.

Depending on the types of seed samples, different types of machine learning models can be used to maximise the efficiency of the vulnerability detection. Numeric and alpha-numeric patterns can be processed using any model. However, for more specific tasks, such as file content mutation and generation of natural language processing or processing of binary data can be used. In literature the most mentioned machine learning models for fuzzing are:

- Graph convolutional networks
- fusion neural networks
- interpretable deep learning models.

4.2.2.3 Code Injection

Merging the technical approaches of fuzzing and password guessing, it is possible to develop an attack framework for code injection. Basic SQL injection attacks have already been implemented by the researchers.[25]

Machine learning will allow to not only use fuzzing but can also incorporate context of the application and the network, that can improve chances for the attack to be successful faster. This way, the less steps the attack requires, the harder it is to prevent using intrusion detection and prevention systems.

Code samples can be used from existing open source code repositories, such as GitHub and GitLab. New datasets keep being published and developed by researchers and a very wide community of machine learning enthusiasts.

Different programming languages have different ways of processing code:

- Code is not executed in the same order as it is written in the source code
- Code is diverse, as it has variables, pointers, functions, and classes, as well as values, that are processed in a different way by a compiler or an interpreter
- Code is contextual, as it can be both malicious and non-malicious in different context.

The key challenge is to pre-process the collected code samples in such a way, that a machine learning model could notice the difference between those samples, even when variables, values and operators change.

[25] Petro, Morris, Weaponizing Machine Learning, DEF CON 25, 2017.

Generative Adversarial networks can be used to expend the existing datasets by generating synthetic samples of code injection queries for a specific programming language.

The model generates new code samples with mutations and checks them with the discriminator and a compiler, to confirm if the syntax of mutations is viable.

4.2.3 Network Traffic Masquerading

To combat modern anomaly detection systems, the malicious actors have capabilities to learn from the target network behaviour passively, and use a network tunnel, that will make malicious traffic behave exactly as non-malicious. The traffic flow has to be recorded and transferred to the system controlled by attackers in order to train the model. The controlled system requires sufficient hardware for the model to be trained.

Network traffic is sequential, and the patterns are time-distributed. Models for sequential and time-distributed data, such as LSTM, should be used for such attack scenarios.

4.2.4 Bots and Botnets

The word bot is a shortened version from "robot". In cyber security, it is a malicious program that acts as a command receiver on a target system, turning one or many such systems into a botnet that is a combination of multiple bots and a Command and Control program. Bots can be distributed by exploiting vulnerabilities, sharing infected files, or sent via spam or phishing emails. The delivery through phishing emails is described in detail in Sect. 4.2.1.

Machine learning can improve the botnet management and general behaviour patterns of bots. Thus, botnet stealth can be greatly enhanced by studying the network traffic of popular services, such as social networks, and replicating the feedback and command signals using similar protocols, ports, and transport paths.

4.2.5 Password Guessing

Passwords have always been a traditional way of access control in computer systems. Password guessing is one of the oldest techniques for gaining access to protected assets.

Dataset for this very specific task can be generated, downloaded online, or scraped from the website or websites of the target organisation.

Generation of the dataset can be done without machine learning using predefined patterns. For example, using a tool called *Crunch*. In combination with a machine learning model, pre-trained on the existing password lists, the number of generated passwords can be greatly reduced, or at sorted by relevance, improving the quality of the dataset and performance of password guessing techniques.

Online datasets can be either collected and generated or used from publicly available data leaks[26] and big hacks. For example, *RockYou2021*[27] is so far the largest known password list and it contains 8.4 billion unique passwords.

Web scraping technique allows developers and attackers to collect every word on the target website and use those words as a password list. Tools such as *cewl* are already available to do it automatically without the need to design custom scripts. Combined with previously mentioned *crunch* and a machine learning model, an intelligent password guessing system can access corporate websites and networks without the need to look for unique vulnerabilities.

PassGAN is a deep learning tool for password guessing, developed by Stevens Institute of Technology in New Jersey, and the New York Institute of Technology. The researchers were able to guess 47% (2,774,269 out of 5,919,936) passwords from RockYou dictionary. The evaluations showed PassGAN was twice as effective as John the Ripper, and equal to HashCat. Furthermore, when combining PassGAN with HashCat the performance was 24% higher than HashCat alone[28]

Hitaj et al., 2019

4.2.6 Ransomware

The only known application of machine learning for ransomware is the ability to hide the decryption keys in the machine learning model or multiple models, as it has already been shown by DeepLocker developers (see Sect. 3.1).

[26] Passwords, https://wiki.skullsecurity.org/index.php/Passwords.

[27] RockYou2021: largest password compilation of all time leaked online with 8.4 billion entries, 7 June 2021, https://cybernews.com/security/rockyou2021-alltime-largest-password-compilation-leaked/.

[28] Hitaj, Briland, Paolo Gasti, Giuseppe Ateniese, and Fernando Perez-Cruz. "Passgan: A deep learning approach for password guessing." In International Conference on Applied Cryptography and Network Security, pp. 217–237. Springer, Cham, 2019.

Ransomware attacks on hospitals

Hospitals in Ireland, New Zealand and Scripps Health in San Diego are reeling from digital extortion attacks

A cyberattack on Ireland's health system has paralyzed the country's health services for a week, cutting off access to patient records, delaying Covid-19 testing, and forcing cancellations of medical appointments

Using ransomware, which is malware that encrypts victims' data until they pay a ransom, the people behind the attack have been holding hostage the data at Ireland's publicly funded health care system, the Health Service Executive. The attack forced the H.S.E. to shut down its entire information technology system

In California, Scripps Health, which operates five hospitals and a number of clinics in San Diego, is still trying to bring its systems back online two weeks after a ransomware attack crippled its data. In New Zealand, a ransomware attack paralyzed multiple hospitals across the country, forcing clinicians to use pen and paper, and postponing nonelective surgeries

Late last year, a ransomware attack on the University of Vermont's Medical Center upended the lives of cancer patients whose chemotherapy treatments had to be delayed or recreated from memory

Ransomware attacks against hospitals surged after two separate efforts—one by the Pentagon's Cyber Command and a separate legal fight by Microsoft—to take down a major botnet, a network of infected computers, called Trickbot, that served as a major conduit for ransomware.[29]

4.2.7 *Cryptomining Malware*

Distributed cryptocurrency mining in cybercrime is second to ransomware. Cybercriminals distribute malware, that uses computational resources of the infected systems for mathematical calculations, required to generate various cryptocurrencies. Computational power of a single consumer-level personal computer is insufficient to generate enough mining power, but a distributed network of such computers can generate sufficient reward without the need to disclose the presence of malware, unlike with ransomware.

Machine learning models can be used for optimising the resource usage in the older hardware for cryptocurrency mining.

---

[29] Nicole Perlroth, Adam Satariano, Irish Hospitals Are Latest to Be Hit by Ransomware Attacks, NY Times, 20 May 2021, https://www.nytimes.com/2021/05/20/technology/ransomware-attack-ireland-hospitals.html.

4.2.8 *Recovery*

Machine learning can enable the less-covered attack methods, such as the recovery of deleted malware or other data.

In this scenario, a malware *A* can analyse the system and spot the signs of data that should have been present, but was removed, for example, malware *B*. Using various approaches, the data can be recovered.

In the file-less mode of operation, malware can write parts of the persistence file to the storage device and remove them, so they could be recovered in the future.

4.2.9 *Cryptanalysis*

Machine learning can be potentially used to grant access to encrypted data, or even optimise the decryption of protected data.

As encryption algorithms are sufficiently resilient, the existing attacks are targeting the implementations of those algorithms, rather than algorithms themselves. Those attacks are known as side-channel attacks. There are three types of side-channel attacks, that can be enhanced using machine learning:

- Profiling-based side-channel attack
- Non-profiling-based side-channel attack
- Profiling-based instruction disassembling template-based attack.

To carry out a profiling-based attack, the malicious actors create models and abstractions that describe the correlation between the side-channel signals and the secret data. Template attacks aim to create a model for each secret key assuming that the accumulated side-channel signals match a specific format or template.

> A deep-learning-based side-channel attack, using the power and [electromagnetic] information across multiple devices has been demonstrated with the potential to break the secret key of a different but identical device in as low as a single trace."[30]
>
> *Golder et al., 2019*

The executed instructions of active programs can be revealed through the analysis of side-channel signals. For example, to identify the executed program block. This method can be used to reverse engineer the programs that are being executed in real time.

[30] Golder, Anupam, Debayan Das, Josef Danial, Santosh Ghosh, Shreyas Sen, and Arijit Raychowdhury. "Practical approaches toward deep-learning-based cross-device power side-channel attack." IEEE Transactions on Very Large Scale Integration (VLSI) Systems 27, no. 12 (2019): 2720–2733.

Machine learning models have already been confirmed to be more effective than the **template attacks** with the limited available information. Supervised machine learning techniques can be used to analyse data distribution, that is required prior to carrying out side-channel attacks.

Alternative to template attacks, machine learning model can be applied to two different stages of **profiling attacks**: the profiling and the attack itself. At the profiling stage the malicious actor isolates encrypted data that can be correlated to a side-channel signal in the cryptographic process. After multiple side-channel signals are collected and correlated with relevant values, the supervised machine learning model is trained. This model will be able to identify similar values for the same device.

> [When] each user has a unique key, the attacker can only observe the traces related to different users. It is assumed that traces that belong to keys with similar Hamming weights are closer to each other. Therefore, collected power traces are grouped into clusters using the clustering algorithm. Next, the attacker predicts the Hamming weight of each cluster. Finally, the secret key is revealed from the predicted Hamming weight by a brute-force attack.
>
> *Elnaggar, Rana, and Krishnendu Chakrabarty. "Machine learning for hardware security: opportunities and risks." Journal of Electronic Testing 34, no. 2 (2018): 183–201.*

Unsupervised machine learning models can be trained to retrieve the secret information without requiring any access to the target system. Unsupervised models can be used to attack elliptic curve multiplications, that is used in public key encryption. Methods like K-means can be used to analyse the signals leaked when single bits are processed serially. To complete the attack, samples with lower probability can be predicted using a guessing attack.

Profiling-Based Instruction Disassembling Template-based technique is another method that allows the attackers to reveal execution instructions of programs during the runtime. However, the method requires data distribution, which changes, and the accuracy of the system is reduced.

Assembly instruction can be extracted using hidden Markov models or similar machine learning models. K-nearest neighbour models can be used to predict blocks of program code in the queue for execution in single-code CPUs without support of advanced architecture (e.g., cache or memory pipelines). Then a neural network can be used to predict the executed instructions and estimate the encrypted values.

4.2.10 Forensics Investigation

Malware can hide from forensic investigation by analysing the behaviour of the system, beyond traditional analysis evasion techniques. Forensics analysis uses imagine software, which makes a copy of the entire system and hardware state. As this process takes time, malware can hypothetically detect the scanning and imagine processes, and migrate from one zone to another, thus remaining undetected.

4.2.11 Attacks Against Hardware

AI can interact with hardware and cyber-physical systems and affects the physical world. To match the AI potential and optimise the computationally intense learning process, hardware had to take a step forward. Autonomous systems rely on the embedded electronic components to perform the computations required for the complex system actions and events. Their purpose, design, mechanical components connected to electronic hardware and a number of linked systems shape the way the software is implemented. Sensors (cameras, radar and acoustic sensors, heat sensors, infrared sensors, etc.) enhance adaptiveness and ensure high performance through a more refined perception of environment enhancing navigation, computer vision, threat detection and other vital functions.

The vast majority of cyber attacks is software-oriented, and only a few existing cyber-attacks can target hardware. Machine learning enables a spectrum of attacks that can target hardware from firmware and from the physical attack vectors.

Firmware is executed with system level privileges. Injection code into firmware would result in the highest level of privileges in the system and potential alterations of physical behaviour of the device.

4.3 Weaponizing AI

The term "weaponized AI" means that the AI system is specifically designed for direct or indirect offense using machine learning. AI-empowered weapons are already a reality, and machine learning techniques have greatly contributed to reinforcement of military capabilities through enhanced vision, language procession, decision-making and control and command, i.e., enabling higher levels of their autonomy.

Autonomous weapon systems (AWS) opened a new era in war technologies, bringing speed of action, enhancing combatting potential, reducing danger to soldiers, providing previously unavailable real-time intelligence from a safe distance, etc. They can operate in the conditions unbearable for humans, communicate wirelessly, provide attack and defence services at the speed exceeding any human capacities while being engaged round the clock. They are limited only by the technological development and become more and more capacitated with its advancement.

The history of the AI-powered weapon systems can be traced back to the 1950s, when acoustic homing torpedoes were first deployed.[31] Almost a century of research and growing investments allowed to develop autonomous machines able to take decisions on battle fields.[32] In late 1980s, the research attempt was undertaken to investigate in the AI controlling a nuclear-powered supersonic aircraft. Another notable approach was to use AI for a real-time monitoring of the nuclear war probability.[33] The "Survival Adaptive Planning Experiment" tested in 1991, was meant to be used in a combination with human intelligence for the use in military operations.

AI systems and automation

[O] lder "first wave" AI systems that employ rule-based decision-making logic have been used in automated and autonomous systems for decades, including in nuclear operations. These expert AI systems use handcrafted knowledge from humans to create a structured set of if–then rules to determine the appropriate action in a given setting. Automated systems of this type are widely used, including in high-consequence operations such as commercial airline autopilots and automation in nuclear power plant operations. Rule-based expert AI systems can often improve reliability and performance when used in predictable settings. However, because such systems can only follow the rules they've been given, they often perform poorly in novel situations or unpredictable environments

Nuclear power plants, commercial airlines, and private space ventures, for instance, all use automation to perform complex operations. Automation also serves niche roles in nuclear operations, including in early warning, targeting, launch control, delivery platforms, and delivery vehicles. Each of these applications, however, relies on mature technology and often retains human control over decision-making prior to the launch of the delivery vehicle.[34]

Horowitz, Scharre & Velez-Green, 2019

---

[31] Barry D. Watts, "Six Decades of Guided Munitions and Battle Networks: Progress and Prospects" (Center for Strategic and Budgetary Assessments, 2007).

[32] It is worth noting that autonomy in machines being of dual-use, a vast research is done for civilian purposes by private business, research centres and students in universities.

[33] Frank Sauer, "Military Applications of Artificial Intelligence: Nuclear Risk Redux", in Vincent Boulanin (ed.), The Impact of Artificial Intelligence on Strategic Stability and Nuclear Risk, SIPRI, Stockholm, 2019.

[34] Michael C. Horowitz, Paul Scharre and Alexander Velez-Green, A Stable Nuclear Future? The Impact of Autonomous Systems and Artificial Intelligence, working paper, December 2019, p. 6–7, https://arxiv.org/ftp/arxiv/papers/1912/1912.05291.pdf.

Autonomy of the weapon systems is not a new phenomenon. It has already existed, e.g. the terminal anti-ballistic defence and point defence systems. Automation has been effectively used in early warning and command and control systems (NC2) warning about threats, e.g., about missiles attacks.[35] AI raised the weapon systems to a completely new level of functionality through granting them "digital intelligence", and subsequently, autonomy in all functions, including in the critical one of engaging the target. Among them are Harpy[36]—loitering munition able to "loiter" over a selected territory awaiting for the target and act without human supervision, point defence systems Phalanx Close-In Weapon System and Patriot, DARPA nuclear-powered supersonic aircraft and autonomous ship "Sea Hunter", etc. The AI-enhancement enables these weapons with the autonomous environmental orientation, computerized vision and face recognition, real-time data collection and analysis, swarming and teaming, immediate decision-making and high speed of its implementation (Scharre, 2018).[37] With the already enhanced hardware of intelligent military machines, it is only the complex decision making that still needs to be perfected to ensure their full autonomy.

The **drone swarming technologies** are currently widely used in surveillance and military operations. Any cyber related issues multiply the swarming drones' danger as they immediately spread through the whole swarm and lead to massive errors, escalating risks.

Drone swarming technology

Unites States

In one of the most significant tests of autonomous systems under development by the Department of Defense, the Strategic Capabilities Office, partnering with Naval Air Systems Command, successfully demonstrated one of the world's largest micro-drone swarms at China Lake, California. The test, conducted in October 2016 and documented on Sunday's CBS News program "60 min", consisted of 103 Perdix drones launched from three F/A-18 Super Hornets. The micro-drones demonstrated advanced swarm behaviors such as collective decision-making, adaptive formation flying, and self-healing.[38]

[35] Air Force Space Command, "Ballistic Missile Early Warning System," March 2017, http://www.afspc.af.mil/About-Us/Fact-Sheets/Display/Article/1126401/ballistic-missile-early-warning-system.

[36] Israel Aerospace Industries, "HARPY: Autonomous Weapon for All Weather", www.iai.co.il/p/harpy.

[37] Paul Scharre, "A Million Mistakes a Second", Foreign Policy, 12 September 2018, available at: https://foreignpolicy.com/2018/09/12/a-million-mistakes-a-second-futureof-war/.

[38] Department of Defence Announces Successful Micro-Drone Demonstration, January 9, 2017, at https://www.defense.gov/Newsroom/Releases/Release/Article/1044811/department-of-defense-announces-successful-micro-drone-demonstration/.

China

At 3:45:39 am on September 20 2020, 3,051 Unmanned Aerial Vehicles took to the skies to break the record for the most Unmanned Aerial Vehicles (UAVs) airborne simultaneously. The benchmark for this record was set by a 2,200 UAV display in Russia on September 4, 2020. This broke the Guinness World Records title set by Intel Corporation (USA) with 2,066 UAVs in Folsom, California, USA on 15 July 2018.[39]

Growing accessibility of cyber tools and hardware already allow to produce rudimentary cyber weapons with low costs and minimum level of knowledge.

4.3.1 Machine Learning for Weapons Autonomy

Since the start of international debates on the threatening consequences of the AI weaponization, a lot of work has been done towards development of new terminology and definitions.[40] One of the critical concepts—autonomy—was put under scrutiny. Scharre and Horowitz (2015) indicate the following three identifiers of autonomy: "human–machine command-and-control relationship, complexity of the machine, type of decision being automated".[41]

The first identifier defines the level of participation of the human operator—from full control to being completely "out of the loop", the level where the full autonomy is exercises. The level of human involvement into the system functionality, a so-called meaningful human control, defines the level of its independence. In a fully AI-enabled autonomy, autonomous systems (devices, platforms) are able to function independently of human operators. However, the level of being semi-autonomous has multiple variations, as well as the debated terms "meaningful human control" and "appropriate levels of human judgment in the use of force".[42] For example, **human control** may vary depending on the system itself, tasks it is implementing, operational

[39] Echo Zhan, 3051 drones create spectacular record-breaking light show in China, 20 October 2020, at https://www.guinnessworldrecords.com/news/commercial/2020/10/3051-drones-create-spectacular-record-breaking-light-show-in-china.

[40] For more reading: https://www.un.org/disarmament/the-convention-on-certain-conventional-weapons/background-on-laws-in-the-ccw/.

[41] Paul Scharre and Michael C. Horowitz, An Introduction to Autonomy in Weapon Systems, Center for a New American Security, February 2015, pp. 6, working paper, February 2015, at https://s3.us-east-1.amazonaws.com/files.cnas.org/documents/Ethical-Autonomy-Working-Paper_021015_v02.pdf?mtime=20160906082257&focal=none.

[42] The J3016 "Levels of Automated Driving" standard issued by the Society of Automotive Engineers (SAE) "defines six levels of driving automation" and considers level 5 to be "full vehicle autonomy". SAE, "SAE Standards News: J3016 Automated-Driving Graphic Update", 7 January 2019, available at: www.sae.org/news/2019/01/sae-updates-j3016-automated-driving-graphic.

environment.[43] Thus, missiles weapons act autonomously within the set time-limits only, while being under human supervision during other functional stages.

The second identifier differentiates automatic devices mechanically responding to commands (e.g., toasters, trip wires, mines), and automated systems—more complex and based on rules (e.g., self-driving cars, programmable thermostats). However, autonomy of computer systems is becoming increasingly harder to define considering the modern advancements in the computational power and complexity of automation.

The third identifier covers the autonomous devices that can self-learn and self-direct based on the collected or received data. Autonomous systems, as compared to the automated ones, have a broader variation in decision making modules and operational field. This concept refers to functions that can be under human control or fully autonomous (e.g., a car may drive on its own but arrive to the destination selected by a human; a weapon system performs some target engagement functions autonomously, but the critical ones are under a human supervision).

Automated versus Autonomous

Understanding the relationships between these terms can be challenging, as they may be used interchangeably in the literature and definitions often conflict with one another. For example, some studies delineate between automated systems and autonomous systems based on the system's complexity, arguing that automated systems are strictly rule-based, while autonomous systems exhibit artificial intelligence. Some, including the Department of Defense, categorize autonomous weapon systems based not on the system's complexity, but rather on the type of function being executed without human intervention (e.g., target selection and engagement). Still others describe AI as a means of automating cognitive tasks, with robotics automating physical tasks. This framework, however, may not be sufficient to describe how AI systems function, as such systems do not merely replicate human cognitive functions and often produce unanticipated outputs. In addition, a robot may be automated or autonomous and may or may not contain an AI algorithm.[44]

Two problems specifically related to automation—automation bias and trust gap—have migrated to the relation between human operators and AI-enabled autonomous devices. Automation is a double-edged sword: overtrusting the machine may lead to mistakes in human judgement, and mistrusting it implies mistrusting the entire concept of autonomy and allowing human mistakes to prevail. The AI-systems have

[43] Daniele Amoroso and Guglielmo Tamburrini, What Makes Human Control over Weapon Systems "Meaningful"?, International Committee for Robot Arms Control, August 2019, available at: www.icrac.net/wp-content/uploads/2019/08/Amoroso-Tamburrini_Human-Control_ICRAC-WP4.pdf.

[44] Daniel S. Hoadley, updated by Kelley M. Sayler, "Artificial Intelligence and National Security", 10 November 2020, Congressional Research Service Report, United States, p. 2, https://fas.org/sgp/crs/natsec/R45178.pdf.

added to the above the lack of full understanding of their operation, risks in identification of the source of commands (environment, operators, malicious programmes) and communication challenges, risks of high-speed response reaction that may be unjustifiable without knowing the context, risks of software failure and vulnerabilities, in general—AI increased the complexity of these already complex systems.[45] The accidents with human deaths reinforced the cautious attitude (e.g., errors in software resulted in the navigation issues of eight F-22 fighter jets,[46] crash of the Air France Flight 447 in 2009,[47] accident with a Tesla car,[48] accident with an Uber car in Arizona[49]).

The US Department of Defense defines Autonomous Weapon System (AWS) as "weapons system that, once activated, can select and engage targets without further intervention by a human operator."[50] Fully autonomous weapons are capable to implement autonomously the full targeting cycles, i.e. find, fix, track, select and engage the target, and do the after-action assessment. The last two functions of selecting and engaging a target have been classified as critical ones (see the ICRC report, 2016)[51] as they are related to taking life and death decisions (iPRAW Report, 2019[52]; Boulanin and Verbruggen, 2017[53]).These capabilities have raised massive

---

[45] Paul Scharre, "Autonomous Weapons and Operational Risk," CNAS Working Paper, February 2016, http://www.cnas.org/sites/default/files/publications-pdf/CNAS_Autonomous-weapons-operational-risk.pdf.

[46] "This Week at War," CNN, 24 February 2007, http://transcripts.cnn.com/TRANSCRIPTS/0702/24/tww.01.html.

[47] "Final Report: On the accidents of 1st June 2009 to the Airbus A330-203 registered F-GZCP operated by Air France flight 447 Rio de Janeiro—Paris," Bureau d'Enquêtes et d'Analyses pour la sécurité de l'aviation civile, English translation, 2012, http://www.bea.aero/docspa/2009/fcp090601.en/pdf/f-cp090601.en.pdf.

[48] The Tesla Team, "A Tragic Loss," Tesla.com, June 30 2016, https://www.tesla.com/blog/tragic-loss; Anjali Singhvi and Karl Russell, "Inside the Self-Driving Tesla Fatal Accident," The New York Times, July 12 2016, https://www.nytimes.com/interactive/2016/2007/2001/business/inside-tesla-accident.html.

[49] Uber's self-driving operator charged over fatal crash, BBC News, 16 September 2020, https://www.bbc.com/news/technology-54175359.

[50] U.S. Department of Defense Directive (DODD) 3000.09, November 21, 2012 Incorporating Change 1, May 8, 2017, https://www.esd.whs.mil/portals/54/documents/dd/issuances/dodd/300009p.pdf.

[51] ICRC, Autonomous Weapon Systems: Implications of Increasing Autonomy in the Critical Functions of Weapons, Geneva, 2016; US Department of Defense (DoD), Directive 3000.09, "Autonomy in Weapon Systems", 2012 (amended 2017); Paul Scharre, Army of None: Autonomous Weapons and the Future of War, W. W. Norton, New York, 2018.

[52] International Panel on the Regulation of Autonomous Weapons (iPRAW), Focus on Human Control, iPRAW Report No. 5, August 2019, available at: www.ipraw.org/wp-content/uploads/2019/08/2019-08-09_iPRAW_HumanControl.pdf.

[53] Vincent Boulanin and Maaike Verbruggen, Mapping the Development of Autonomy in Weapon Systems, Stockholm International Peace Research Institute (SIPRI), Stockholm, 2017, available at: www.sipri.org/sites/default/files/2017-11/siprireport_mapping_the_development_of_autonomy_in_weapon_systems_1117_0.pdf.

protests of technical experts, researchers and developers, producers, scientists, opinion leaders.[54]

During the implementation process, the machine learning-based systems (trained software) take decisions based on the trained capacities, or/and data collected from environment through sensors, and enables hardware for action. The software receives data from sensors and takes decisions on tracking, reporting, and executing commands. The loosely simplistic model of an AWS would consist of the following key components—sensors, an object recognition software, and a physical device that executes any action. It can also include a smart security camera, a self-driving or aerial vehicle, or a smart light switch with motion detection. All those devices have sensors, object recognition and motion recognition are present, and a physical output ensures the execution of the mechanical tasks, such as moving the position of the device, or turning on or off a certain function.

AWSs at the technical level do not have any electronic components that make them inherently malicious or offensive. Which implies that, from the ethical point of view, a mounted or attached weapon matters more than the software controlling it.

4.3.2 AWS Vulnerabilities

Enabled by software and receiving information from sensors, AWSs reveal multiple vulnerabilities. They can become dysfunctional due to technical issues, and the control over AWSs may be acquired by malicious actors though cyber attacks (Abaimov and Ingram, 2017).[55] Being vulnerable to cyber attacks, AWS risk to be manipulated or even fail during operation.[56] Though the access to such powerful military equipment is strictly secured, there is never a full guarantee against an intrusion. And the wider use of autonomous systems will entail increasing threats (Danks, 2020[57]).

AWS entailing increased attack surface

From a cyber threat perspective, the proliferation of autonomous systems and devices is expected to increase the attack surface available to adversaries and malicious actors. For example, autonomous weapons systems that include

[54] Campaign to Stop Killer Robots, https://www.stopkillerrobots.org/.

[55] See Stanislav Abaimov and Paul Ingram, Hacking UK Trident: A Growing Threat, British American Security Information Council, June 2017.

[56] Michał Klincewicz, "Autonomous Weapons Systems, the Frame Problem and Computer Security", Journal of Military Ethics, Vol. 14, No. 2, 2015.

[57] Danks, D. 2020 How Adversarial Attacks Could Destabilize Military AI Systems. Available from: https://spectrum.ieee.org/automaton/artificial-intelligence/embedded-ai/adversarial-attacks-and-ai-systems [Accessed 23rd September 2020].

> a tether, enabling the remote control of a system from a supplying country wishing to ensure compliance of the use of its systems with international humanitarian law, could result in the embedding of back doors and kill switches limiting the value of autonomous system assets and potentially making them vulnerable to disruption or manipulation by other third parties. Similarly, the use of autonomous vehicles for logistics could be targeted by adversaries leveraging cyber vulnerabilities or adversarial AI to disrupt the logistics and supply chains of a military operation.[58]
>
> *Cyber Threats and NATO 2030: Horizon Scanning and Analysis', 2020*

Numerous scenarios have been developed to demonstrate vulnerability of AWS stemming from the AI technologies. They include cyber threats, e.g., poisoning algorithms at the initial stage of machine learning, or during maintenance; or cyber-physical threats when the cyber agents infect AI through external devices, etc.Unintentional human errors at the development stage of complex systems are inevitable by themselves,[59] in addition, malicious programmes can be introduced at all stages of the systems production, i.e., design, development, assembling, maintenance.[60] The systems re-training and update may bring additional errors.

GPS spoofing is an example of a cyber attack with a reported case of a successful hijack of an unmanned aerial vehicle in Iran in 2011.[61] Image recognition can be disoriented through a number of methods to bypass or avoid entirely any facial recognition technology, which includes make up, face facial hair, face coverings, and 3D-printed masks of other people.[62]

As discussed previously, the way some algorithms arrive to decisions, is still a "black box", meaning that the dysfunctionality may not be understood and may require the full circle of a new model creation and training. Communication of autonomously functioning algorithms may result in unforeseen clashes, caused by software bugs. It may take time to discover where the error of a false command come

---

[58] Cyber Threats and NATO 2030: Horizon Scanning and Analysis', ed. by A. Ertan, K. Floyd, P. Pernik, Tim Stevens, CCDCOE, NATO Cooperative Cyber Defence Centre of Excellence, King's College London, Williams and Mary, 2020, p. 92, at https://ccdcoe.org/uploads/2020/12/Cyber-Threats-and-NATO-2030_Horizon-Scanning-and-Analysis.pdf.

[59] John Borrie, Security, Unintentional Risk, and System Accidents, United Nations Institute for Disarmament Research (UNIDIR), Geneva, 15 April 2016, available at: https://tinyurl.com/yyaugayk; P. Scharre, Autonomous Weapons and Operational Risk.

[60] Ivan Evtimov et al., "Robust Physical-World Attacks on Deep Learning Models", Proceedings of the IEEE Conference on Computer Vision and Pattern Recognition, 2017, available at: https://arxiv.org/pdf/1707.08,945.pdf.

[61] Sydney J. Friedberg, "Drones Need Secure Datalinks to Survive vs. Iran, China", Breaking Defense, 10 August 2012, at http://breakingdefense.com/2012/08/drones-need-secure-datalinks-to-survivevs-iran-china/.

[62] "Deep Neural Networks Are Easily Fooled: High Confidence Predictions for Unrecognizable Images", Proceedings of the IEEE Conference on Computer Vision and Pattern Recognition, 2015, pp. 427–436.

from, or, even worse, in the absence of a proper human control, the dysfunctionality may be unnoticed for a long period of time.This can refer to false alarms and military escalation.

The issue of creating false alarms has been discussed with regards to the intrusion detection systems (IDS). The same challenge exists in AWSs operating in the hostile environment and is especially dangerous in case of swarms when the error propagates simultaneously and with a high speed.

The Dubai Airshow, 2019

At the Dubai Airshow in 2019, the chief of staff of the US Air Force, General David Goldfein, presented the simulated engagement of an enemy navy vessel with a next-to-fully automated kill chain. The vessel was first picked up by a satellite, then target data was relayed to airborne surveillance as well as command and control assets. A US Navy destroyer was then tasked with firing a missile, the only remaining point at which this targeting cycle involved a human decision, with the rest of the "kill chain ... completed machine to machine, at the speed of light.[63]

Growing threats from the AI-powered malware, AWSs, increases the "non-nuclear threats to nuclear weapons and their associated command, control, communication, and information (C3I) systems".[64]

The "brittleness" inherent to emerging forms of automation, along with omnipresent risk of automation bias, human–machine interaction failures, and unanticipated machine behavior, all potentially limit the roles that automation can safely fill.[65]

---

[63] Frank Sauer, Stepping back from the brink: Why multilateral regulation of autonomy in weapons systems is difficult, yet imperative and feasible, International Review of the Red Cross, March 2021, https://international-review.icrc.org/articles/stepping-back-from-brink-regulation-of-autonomous-weapons-systems-913. The video with the speech Video: Here's How the US Air Force Is Automating the Future Kill Chain", Defense News, 2019, at www.defensenews.com/video/2019/11/16/heres-how-the-us-air-force-is-automating-the-future-killchain-dubai-airshow-2019/.

[64] James M. Acton (ed.), Entanglement: Chinese and Russian Perspectives on Non-Nuclear Weapons and Nuclear Risks, Carnegie Endowment for International Peace, 2017, p. 1, at: http://carnegieendowment.org/files/Entanglement_interior_FNL.pdf.

[65] Michael C. Horowitz, Paul Scharre and Alexander Velez-Green, A Stable Nuclear Future? The Impact of Autonomous Systems and Artificial Intelligence, working paper, December 2019, p. 8, https://arxiv.org/ftp/arxiv/papers/1912/1912.05291.pdf.

Discussions are ongoing about improving the monitoring of the **nuclear command and control** and early warnings with AI-systems.[66] However, the human presence is always considered as a strong necessity in this high-risk environment.[67]

With all the instability and uncertainty of AI, there are two reason to continue the research in the area of autonomy: mitigation of false alarms through human–machine collaboration, and deterrence against the threat of repeated attacks against the state.

4.4 Conclusion

This chapter has provided the analysis of the machine-learning-based attacks and AI use for offensive actions, including the weaponized AI as its most dangerous implementation. A machine learning model can be an attack tool and a target.

Machine learning can be used to increase the evasive capabilities of malware, success rate of phishing emails, and efficiency of conventional attack tools. It also provides a set of intrinsically new attack vectors, such as data poisoning for misclassification. Machine learning also creates a vulnerability that can become an entry point for the attackers.

The findings show a high potential of AI in enhancing the existing malicious cyber arms and developing a new more elusive malware. The AI equally granted a higher level of autonomy to weapons and reinforced their potential with computer vision, orientation in the complex environment, real time data procession, high-speed round the clock offensive actions. Fully Autonomous Weapon Systems (AWSs), capable of implementing the full targeting cycle independently, including the decision making on the targets engagement, are inhumane, and dangerous not only because of being weapons, but also because of multiple vulnerabilities stemming from the machine learning itself, and added during the life-cycle of AI-systems. Unless the cyber security of AWSs is properly ensured, their vulnerability to cyber attacks and risks of manipulation lead to more harm than benefits from their use.

---

[66] Philip Reiner and Alexa Wehsner, "The Real Value of Artificial Intelligence in Nuclear Command and Control", War on the Rocks, 4 November 2019, at: https://warontherocks.com/2019/11/the-real-value-of-artificial-intelligence-in-nuclearcommand-and-control/.

[67] Edward Geist and Andrew J. Lohn, How Might Artificial Intelligence Affect the Risk of Nuclear War?, RAND Corporation, 2018, at www.rand.org/content/dam/rand/pubs/perspectives/PE200/PE296/RAND_PE296.pdf; Vincent Boulanin, Lora Saalman, Petr Topychkanov, Fei Su and Moa Peldán Carlsson, Artificial Intelligence, Strategic Stability and Nuclear Risk, SIPRI, Stockholm, June 2020, at: www.sipri.org/sites/default/files/2020-06/artificial_intelligence_strategic_stability_and_nuclear_risk.pdf.

The next chapter will discuss the international resonance as a reaction to the discussed evolving threats and highlight the concerted international efforts in ensuring the benefits of AI are used for peaceful purposes, in a stable and safe digital environment.

Reference

1. Abaimov S, Martellini M (2020) Cyber arms: security in cyberspace. CRC Press

Chapter 5
International Resonance

AI implementation is accompanied by heated international discussions of its potential as a new technology, impact on society, incorporation into legal and ethical environments, etc. AI technologies have brought progressive changes, capabilities to resolve issues unresolvable previously, e.g., environmental and health challenges, assistance to people with disabilities. Increasing the speed of data procession, bringing openness and interconnectedness, they have also magnified all existing challenges in political, economic and social areas, making them more evident and requiring immediate solutions.

> **Connectivity is the new geopolitics:** If, in 2005, there were around one billion Internet users worldwide, today that number stands at almost four billion and rising. At the same time, the number of connected devices is increasing exponentially, powered by the fast-growing Internet of Things and the Fourth Industrial Revolution. Indeed, 'connectivity' is becoming a forceful expression of political power and global ambition, far surpassing mere economics.[1]
>
> *ESPAS Global Trends to 2030: Challenges and Choices for Europe, 2019*

With full understanding of the benefits, mindful optimism and consideration of opportunities, the international community raises global concerns over emerging threats from the AI-empowered machines and lack of internationally agreed norms, regulations and standards. Multiple reports published by research agencies and security organisations in the past decades warn humanity about the risks accompanying these inventions.

[1] Florence Gaub et al., ESPAS Global Trends to 2030: Challenges and Choices for Europe, European Policy and Analysis System, An Inter-institutional research project, Report, 2019, p. 3. https://espas.secure.europarl.europa.eu/orbis/sites/default/files/generated/document/en/ESPAS_Report2019.pdf.

S. Abaimov and M. Martellini, *Machine Learning for Cyber Agents*,
Advanced Sciences and Technologies for Security Applications,
https://doi.org/10.1007/978-3-030-91585-8_5

The malicious use of AI could threaten **digital security** (e.g., through criminals training machines to hack or socially engineered victims at human or superhuman levels of performance), **physical security** (e.g., non-state actors weaponizing consumer drones), and **political security** (e.g., through privacy-eliminating surveillance, profiling, and repression, or through automated and targeted disinformation campaigns).[2]

The Malicious Use of Artificial Intelligence: Forecasting, Prevention, and Mitigation, Report, 2018

This chapter will review the international debates and actions over the AI application in civil life and for military purposes, analysing more in detail the growing threat that may bring an existential danger. It will also highlight the latest multilateral initiatives and agreements as per the outcomes of global forums and appeals of the progressive-minded world leaders.

5.1 Debates Over AI Integration and Governance

Machine learning is a relatively new technology. As already mentioned, it is different from a conventional precise logic based on the linear equations and conditional "if..then..else" statements. The uncertainty in the complex behaviour of such technology causes multiple questions about its implementation. Among the questions most widely discussed around the AI technologies at the national and international levels are reliability of machine learning techniques, transparency in application of AI algorithms, use of correct and representative data, privacy regulations, accountability for decisions taken by machines and responsibility for errors, compliance with legal regulations, respect of human rights and democratic principles, ethical and moral standards.

For ease of analysis, we will group the debated topics as follows:

- technical issues
- ethical and legal issues
- governance
- military use of AI-enabled systems.

The identified groups are not mutually exclusive, as they overlap and reinforce each other in multiple areas, making the debates on AI development and adoption—a complex argument at national, international, and global levels.

---

[2] Report on The Malicious use of artificial Intelligence: Forecasting, Prevention and Mitigation, February 2018, last retrieved 15 July 2018 at http://img1.wsimg.com/blobby/go/3d82daa4-97fe-4096-9c6b-376b92c619de/downloads/1c6q2kc4v_50335.pdf. Among the contributors to the report are academics and researchers from the Oxford University, Cambridge University, Stanford University, the Electronic Frontier Foundation, artificial intelligence research group OpenAI, etc.

Debates over **technical issues** include discussions around innovations, research and development of technologies, their safety and security, standardization. The discussions are organised through professional societies and associations, workshops, conferences, high-level working groups, research conferences and academic publications, etc.

Legal and ethical issues are discussed with relation to creating favourable environment for accommodation and integration of the new technologies into our life, at the same time protecting humans from any potential threats. Responsibility for decisions taken by machines, their transparency and accountability, fairness in data representation and reliability of outcomes, integration of human values and morality in the machine decision making process are broadly discussed in this area.

Governance of the emerging technologies includes authoritative control over their integration into social, economic, political life, ensuring compliance with policies and procedures, minimisation of risks without creating unnecessary barriers for innovations. Investments into research, development of multisectoral plans of action, re-education of people, whose jobs are affected by the AI-technologies, can be also included into this group.

Military use of AI has created broad "whole-of-society" debates over the AI offensive capabilities and deserves special attention. The discussion includes deliberations over autonomous weapon systems (AWS), and appeals to ban the lethal AWS (LAWS).

Considering the diversity of legal systems, national specificities and cultural differences, but equally recognizing the growing global interconnectedness, governments, international organizations, academia, representatives of industries, producers and developers, communities and opinion leaders join their forces to agree and implement internationally acceptable solutions.

5.1.1 Debates Over Technical Issues

Researchers and developers seek joint solutions in advancing the technological progress, increasing performance and efficiency of innovative solutions. The list of tasks to address includes but is not limited to:

- establishing and agreeing the universal terminology
- developing technical guidance and disseminating best practices
- creating and making available high-quality data
- developing more reliable and explainable algorithms
- responding to safety and security challenges both for computer systems and information
- introducing internationally recognised educational programs, certificates and licenses
- raising awareness on challenges, threats, and risks
- organising simulation exercises to ensure readiness for emergencies.

As broadly discussed in the previous chapters, the very implementation of the AI technologies at this stage represents a considerable challenge. First and foremost, due to their own immaturity and specific vulnerabilities, requiring enhanced safety and security measures. Global debates with participation of the world leading scientists are addressed to the whole technical community, so the technology can be jointly taken to the new frontier.

Some global debates

The 2019 Montreal AI Debate between Yoshua Bengio and Gary Marcus included the discussion about AI development status and prospects.

The AAAI-2020 discussion with Daniel Kahneman (Nobel Laureate in economics) and Geoffrey Hinton, Yoshua Bengio and Yann LeCun, the 2018 Turing Award recipients, touched the prospects (neurons and symbols in AI, distributed and localist representations, taxonomy for neurosymbolic AI proposed by Henry Kautz), new directions for research in explainable AI, and raised multiple concerns. Discussion of systems based on deep learning and symbolic reasoning.

With AI research is gradually becoming more and more multidisciplinary, these technical discussions involve not only computer specialists, but also experts from various scientific fields—electronic engineers, physicists, neuroscientists, psychologists, philosophers, and diplomats. Computer specialists and engineers design, shape and test the new technologies, linguists analyse and unify the terminology definitions, neuroscientists enrich the AI capabilities copying the work of human neurons to machines, and learning about the work of human brain from the machine models, physicists propose visionary ideas from quantum mechanics and the next generation of hardware platforms.

But all this comes at a cost. Errors in production have a global impact. Not all experiments are successful, but these are exactly their failures that bring the whole world attention to specific AI vulnerabilities and urge to be cautious.

AI interacting with humans

On March 24, 2016 Microsoft introduced Tay, an AI-based chat robot, to Twitter. The company had to remove it only 16 hours later. It was supposed to become increasingly smarter as it interacted with humans. Instead, it quickly became an evil, Hitler-loving, Holocaust-denying… "Bush did 9/11"-proclaiming chatterbox. Why? Because it worked no better than kitchen paper, absorbing and being shaped by the tricky and nasty messages sent to it. **Microsoft had to apologise**.

https://www.sciencefocus.com/future-technology/should-we-be-worried-about-ai/

There exist a number of international professional associations and societies that organize discussions, conferences, participate in advisory committees to the respective international bodies and governments. Scientific journals publish papers that pass a thorough peer review of independent experts from different countries exchanging opinions and disseminating knowledge. For example, the IEEE—one of the "the world's largest technical professional organization dedicated to advancing technology for the benefit of humanity."[3] The international standards development organizations include the International Telecommunications Union (ITU), ISO/IEC, IEEE, and many others. Their contribution to accumulating, registering, measuring and organizing new knowledge is remarkable.[4]

Professional associations and societies

Standardization

- International Standards Organization (ISO/IEC)
- International Telecommunication Unit (ITU)
- Institute of Electrical and Electronics Engineers (IEEE)
- National Institute of Standards and Technology (NIST)

Scientific societies for AI research

- American Association for Artificial Intelligence (AAAI)
- European Coordinating Committee for Artificial Intelligence (ECCAI)
- Society for Artificial Intelligence and Simulation of Behavior (AISB)
- Association for Computing Machinery (ACM) with a group on artificial intelligence SIGART

Conferences

- The International Joint Conference on AI (IJCAI)[5]
- The AAAI Conference on Artificial Intelligence[6]

[3] Institute of Electrical and Electronics Engineers, IEEE, https://www.ieee.org/.

[4] Artificial Intelligence and Ethics in Design: Responsible Innovation, IEEE educational activities, https://innovationatwork.ieee.org/AI_Ethics/.

[5] The International Joint Conference on Artificial Intelligence: Workshop on Explainable Artificial Intelligence (XAI), https://www.ijcai.org/.

[6] The AAAI Conference on Artificial Intelligence, https://www.aaai.org/Conferences/AAAI/aaai.php.

Major journals

- Electronic Transactions on Artificial Intelligence[7]
- Artificial Intelligence[8]
- Journal of Artificial Intelligence Research[9]
- IEEE Transactions on Pattern Analysis and Machine Intelligence[10] (TPAMI)

Professional education also becomes more international, universally standardised, and now covers multidisciplinary studies. At the national level, general trainings are organized for the public to raise awareness about cyber security threats, and for the certification of the designated personnel, e.g. lawyers, senior managers. Cyber experts pass specialized trainings while preparing for exams for the **internationally recognized certificates** that raise their level of knowledge but also contributes for career development. For examples, training for the Cloud Security knowledge (Cloud Security Alliance), CISSP (ISC^2), OSCP (Offensive Security), CEH (EC-Council), Security+ (CompTIA), etc.

To ensure the responsible and safe use of the new technology, **technical standards** are being developed and **technical guidance** introduced not only for field experts, but also for non-technical users world-wide. The latter audience is more numerous, and consequently requires their adaptation to be able to understand how they work and take informed decisions. The debatable issue here is related to the differences in the national approaches, and technical standards acceptable in one country may not necessarily be allowed in some others. With the spread of IoTs globally, the minimum requirements—safety conditions—and a certain level of controls should be ensured. While developing the standards it is essential to also ensure the agreed technical criteria for the **international monitoring and evaluation**[11] of their application, quality of data bases, reporting of security breaches by the national focal points. This will reinforce the governance frameworks.

---

[7] Electronic Transactions on Artificial Intelligence (ETAI), https://www.ida.liu.se/ext/etai/.

[8] Artificial Intelligence, An International Journal, https://www.journals.elsevier.com/artificial-intelligence.

[9] Journal of Artificial Intelligence Research, JAIR, https://www.jair.org/index.php/jair.

[10] IEEE Transactions on Pattern Analysis and Machine Intelligence, IEEE, https://ieeexplore.ieee.org/xpl/RecentIssue.jsp?punumber=34.

[11] International Electrotechnical Commission. Functional Safety and IEC 61,508; Redrup Y. 2018. Google to make AI accessible to all businesses with Cloud AutoML. Australian Financial Review.

Recommendations from technical experts

1. Policymakers should collaborate closely with technical researchers to investigate, prevent, and mitigate potential malicious uses of AI.
2. Researchers and engineers in artificial intelligence should take the dual-use nature of their work seriously, allowing misuse-related considerations to influence research priorities and norms, and proactively reaching out to relevant actors when harmful applications are foreseeable.
3. Best practices should be identified in research areas with more mature methods for addressing dual-use concerns, such as computer security, and imported where applicable to the case of AI.
4. Actively seek to expand the range of stakeholders and domain experts involved in discussions of these challenges.[12]

The Malicious Use of Artificial Intelligence: Forecasting, Prevention, and Mitigation, Report, 2018

With the overall accessibility of equipment, and decreasing costs, AI-enhanced devices may be produced even with the minimal knowledge, also in an unpredictable way. Thus, to ensure security at the production level and avoid malicious use of the new technology, **international educational certificates and licences** should be introduced for people developing and producing AI goods. This will also reinforce the responsibility of producers and contribute to strengthening the legal environment.

5.1.2 Debates Over Legal and Ethical Issues

The new omnipresent technologies that are able to trespass the national boundaries require a reinforced legal environment for their accommodation and integration. Technological benefits should bring advantages to all humanity without any discrimination, and all should be protected from threats. These debates are organized around transparency, reliability and accountability for decisions taken by machines, trust in machines predictions, contestability of decisions, data and privacy protection, legal responsibility for errors and reputational damages in case of leakages; human rights issues and elimination of discrimination; democratic principles of fairness and inclusiveness, protection from social manipulations, etc.

Recognizing enormous potential to improve the society, and accompanying risks, the world expects a safe and reliable AI used within the ethical frameworks and appropriate legal environment—regulations establishing responsibility, accountability and

[12] Miles Brundage et al., The Malicious Use of Artificial Intelligence: Forecasting, Prevention, and Mitigation, February 2018, https://arxiv.org/abs/1802.07228.

liability for all kinds of damages, including for social manipulations, data theft, privacy violations and other related issues.

AI influencing people's choices

Using the recurrent neural networks and deep reinforcement learning,[13] "[s]cientists ran three experiments where participants played games against a computer. As the machine gained insights from the behaviour underlying participant responses, it identified and targeted vulnerabilities in people's decision-making to steer them towards particular actions or goals.[...].

The research was conducted in partnership with the Australian National University (ANU), Germany's University of Tübingen, and Germany's Max Planck Institute for Biological Cybernetics."[14]

Commonwealth Scientific and Industrial Research Organisation (CSIRO), Australia's national science agency

One of the broadly debateable questions is the responsibility for failures. In the long chain of the people behind the implemented AI-device—data scientists, designers, developers, operators, supervisors, owners—who exactly is responsible for potential dysfunctionalities? The **responsibility gap** is evident. The lack of understanding behind the functions of algorithms, control for decisions taken by machines shows that people, e.g. surgeons, cannot be always responsible. However, the common sense advises that people should always bear responsibility, as in case of inability to control they should stop developing these technologies.

Google about responsibility

it is not appropriate for moral or legal responsibility to be shifted to a machine. No matter how complex the AI system, it must be persons or organizations who are ultimately responsible for the actions of AI systems within their design or control.[15]

Google, Perspectives on issues in AI governance, 2019

---

[13] Amir Dezfouli, Richard Nock, Peter Dayan, Adversarial vulnerabilities of human decision-making, Proceedings of the National Academy of Sciences, November 2020, 117 (46) 29,221–29,228; https://doi.org/10.1073/pnas.2016921117.

[14] AI's influence on human decision-making: new CSIRO research, Commonwealth Scientific and Industrial Research Organisation, Australian Government, 10 February 2021, https://www.csiro.au/en/news/News-releases/2021/AIs-influence-on-human-decision-making-new-CSIRO-research.

[15] Google, Perspectives on issues in AI governance, 2019, p. 26. https://ai.google/static/documents/perspectives-on-issues-in-ai-governance.pdf.

EU Committee on Legal Affairs

[A]utonomous decision-making should not absolve humans from responsibility, and … people must always have ultimate responsibility for decision-making processes so that the human responsible for the decision can be identified.[16]

The issue of accountability for AI decisions is being raised at both technical and ethical levels. Multiple questions are being asked with relation to our capacity to respond to increased information flow, violation of privacy, job losses. Debates are ongoing, and it remains the people's choice how to address the responsibility gap.

Strategic questions on shaping the digital eco-system

For Europe, a set of policy questions arises from this development: do we want to become the shapers of digital ethics, and contribute a distinctly European, carefully calibrated and agile regulatory digital eco-system to the ongoing race for digital leadership?

Are we willing to invest, de-regulate, anticipate, and legislate?

Do we develop a digital strategy to address all the socioeconomic issues resulting from digital innovation? Do we try and pre-empt the disruption of the labour market by re-training those whose jobs are at risk of automation?

Do we counter the development of technologies that undermine democracy and human rights both in Europe and abroad?

Do we teach ourselves how to handle the overflow of information and life assistance these new technologies can provide?

Are we equipped to deal with hyper-progress at both the individual, as well as collective level?.[17]

ESPAS Global Trends to 2030: Challenges and Choices for Europe, 2019

Data protection and **safety** require protecting privacy, information itself, and computer systems and networks from cyber attacks and damage. It is the area that is closely linked with the technical side. Rössler [1][18] defines violations of privacy

[16] Report on artificial intelligence: questions of interpretation and application of international law in so far as the EU is affected in the areas of civil and military uses and of state authority outside the scope of criminal justice (2020/2013(INI)). Committee on Legal Affairs Rapporteur: Gilles Lebreton, 4.1.2021, p. 6, https://www.europarl.europa.eu/doceo/document/A-9-2021-0001_EN.pdf.

[17] Florence Gaub et al., ESPAS Global Trends to 2030: Challenges and Choices for Europe, European Policy and Analysis System, An Inter-institutional research project, Report, 2019, p. 36. https://espas.secure.europarl.europa.eu/orbis/sites/default/files/generated/document/en/ESPAS_Report2019.pdf).

[18] Rössler [1], p. 9.

as "illicit interference in one's actions, as illicit surveillance, as illicit intrusions in rooms or dwellings." This definition is becoming more relevant with the growing interconnectivity, and especially to the machine learning that requires big amounts of data.

Artificial intelligence risks to privacy demand urgent action

Artificial intelligence can be a force for good, helping societies overcome some of the great challenges of our times. But AI technologies can have negative, even catastrophic, effects if they are used without sufficient regard to how they affect people's human rights.[19]

Michelle Bachelet, UN High Commissioner for Human Rights

The widespread IoT allows to get information from home and wellness appliances, governments exercise national surveillance, monitor social behaviour. As there are different national standards in relation to the issue of privacy, the development of internationally agreed norms is also a challenge. Moreover, the national privacy acts may be lagging behind the technological advantages, e.g. the Cambridge Analytica scandal. Biometrics currently includes not only fingerprints, but also facial and voice recognition.

Ethical questions for AI systems

Ethical questions regarding Al Systems pertain to all stages of the Al System life cycle, understood to range from research, design, and development to deployment and use, including maintenance, operation, trade, financing, monitoring and evaluation, validation, end-of-use, disassembly, and termination.

In addition, **Al actors** can be defined as any actor involved in at least one stage of the Al life cycle, and can refer both to natural and legal persons, such as researchers, programmers, engineers, data scientists, **end-users**, large technology companies, small and medium enterprises, start-ups, universities, public entities, among others.[20]

UNESCO

[19] Artificial intelligence risks to privacy demand urgent action—Bachelet, https://www.ohchr.org/RU/NewsEvents/Pages/DisplayNews.aspx?NewsID=27469&LangID=E.

[20] Preliminary Report on the first draft of the Recommendation on the Ethics of Artificial Intelligence.

https://unesdoc.unesco.org/ark:/48223/pf0000374266/PDF/374266eng.pdf.multi.

In case of any gaps in the data representation, this may lead to **discrimination** in decisions, replicating existing structural inequalities, for example with regards to under sampled minorities. This may also lead to the wrong **predictions**. How far can we trust machines in predicting future development if their decisions maybe taken from the incorrectly compiled data?[21]

Recently, the world has witnessed a surge in international initiatives for the ethical AI.

Vatican signs for the ethical AI

The Pontifical Academy for Life, Microsoft, IBM, FAO, the Italian Ministry of Innovation (part of the Italian Government), signed as first the "**Call for an AI Ethics**", a document developed to support an ethical approach to Artificial Intelligence and promote a sense of responsibility among organizations, governments, institutions and the private sector with the aim to create a future in which digital innovation and technological progress serve human genius and creativity and not their gradual replacement.[22]

The document includes six principles:

Transparency: In principle, AI systems should be explainable.

Inclusion: The needs of all human beings must be taken into consideration, so that all can benefit from and enjoy the best possible conditions to express themselves and grow.

Responsibility: Designers and developers of artificial intelligence solutions must act responsibly and transparently.

Impartiality: systems should not be created or operated according to bias, in view to protect human equality and dignity.

Reliability: artificial intelligence systems must be able to operate reliably.

Security and privacy: artificial intelligence systems must work securely and respect the privacy of users.

Private business, developing and producing technology companies, have joined their forces with the civil society representatives, academics, researchers, in various forms of collaboration. "Partnership on AI" was created with the aim "to study and formulate best practices on AI technologies, to advance the public's understanding of AI, and to serve as an open platform for discussion and engagement about AI and its influences on people and society".[23] It has more than 100 partners from more than 13 countries, and among its members are Adobe, Amazon, Apple, Baidu, BBC, Chatham House, Facebook, Google, IBM, Intel, Microsoft, Sony, United Nations Development Programme (UNDP), etc.

---

[21] Moses L B, Chan J. 2016. Algorithmic prediction in policing: assumptions, evaluation, and accountability. Policing and Society, 28(7): 806–822.

[22] Vatican signs for the ethical AI, 28 February, 2020, https://www.romecall.org/.

[23] Partnership on AI, PAI, https://www.partnershiponai.org/about/.

Safety–critical AI

Effective and careful applications of pattern recognition, automated decision making, and robotic systems show promise for enhancing the quality of life and preventing thousands of needless deaths. However, where AI tools are used to supplement or replace human decision-making, we must be sure that they are safe, trustworthy, and aligned with the ethics and preferences of people who are influenced by their actions.[24]

Partnership on AI

Democratic principles require **fairness** in decisions, **equality, equity, inclusiveness, non-discrimination**. How far can we trust AI in ensuring all of them? We may argue that the machines will take decisions based on the provided data and set parameters. But humanity itself accumulated lots of unresolved ethical and moral issues related to specifically difficult choices. How can we expect them to be resolved by machines? With AI affecting job markets, how can we ensure that the people whose jobs were replaced by machines will get equal opportunities for further employment? How can we ensure that the governments do not use AI for their own benefits ignoring the interest of citizens? These social responsibility issues are of the legal, ethical and governance nature at the same time. It may be that the current debates are too much concentrated around technologies, avoiding heavy criticism of ethics and morality of the modern society itself.

On the trust

In Artificial Intelligence, trust is a must, not a nice to have. With these landmark rules, the EU is spearheading the development of new global norms to make sure AI can be trusted. By setting the standards, we can pave the way to ethical technology worldwide and ensure that the EU remains competitive along the way. Future-proof and innovation-friendly, our rules will intervene where strictly needed: when the safety and fundamental rights of EU citizens are at stake.[25]

Margrethe Vestager, Executive Vice-President for a Europe fit for the Digital Age

[24] Partnership on AI, https://www.partnershiponai.org/about/.

[25] Europe fit for digital age: Commission proposes new rules and actions for excellence and trust in Artificial Intelligence, Press release, 21 April 2021, https://ec.europa.eu/commission/presscorner/detail/en/ip_21_1682.

Is our society ready to accommodate powerful technologies? How shall we find sustainable harmonized solutions and build consensus between the heterogeneous political systems: democracy, authoritarian regimes, conflict and post-conflict zones? Will the voice of researchers, designers and developers be heard by technological giants? We can argue that we are appealing to human values, common benefits, achievements of the multilateral cooperation. Still we need to recognize the diversity in understanding privacy, personal freedoms, levels of control over population related to the different political systems, prioritization of profit gains. Here, the starting point of discussing basic safety and security may further allow more extensive dialogue supported by civil society activists.

Transparency in decision making requires that the people know that the decision was taken by the machine and how it was taken, also the limitations accompanying the methods used. In the case of the decisions taken by machines, who is legally **accountable?** How can we ensure accountability and transparency against algorithmic decisions?[26] With the need to provide high-quality and representative data, how shall we treat business secrets and confidential information?

World Health Organization: guiding recommendations for promoting transparency in AI for health

Introduction of any AI technology must be sufficiently transparent that it can be criticized, by the public or by internal review mechanisms. Specific considerations:

- The source code should be fully disclosed.
- Algorithms must be open to criticism by an in-house or other appropriate expert.
- The data used to train the algorithm, whether certain groups were systematically excluded from such data, how the training data were labelled and by whom (including expertise and appropriateness of labelling) should be known.
- The underlying principles and value sets used for decision trees should be transparent.
- The learned code should be available for independent audit and review by appropriate third parties.[27]

Ethics & Governance of Artificial Intelligence for Health, 2021

[26] Reisman D, Schultz J, Crawford K et al. 2018. Algorithmic Impact Assessments: A Practical Framework for Public Agency Accountability; Kleinman Z. IBM launches tool aimed at detecting AI bias.

[27] World Health Organization, Ethics & Governance of Artificial Intelligence for Health, Guidance, 28 June 2021, p. 146, https://www.who.int/publications/i/item/9789240029200.

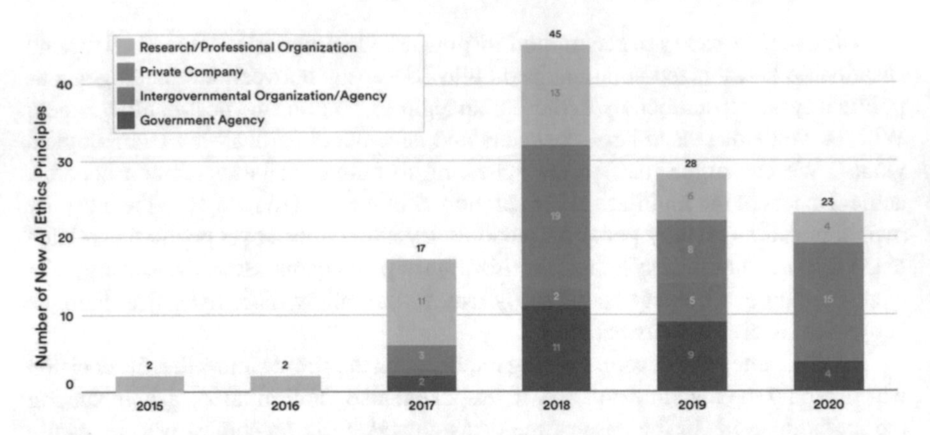

Fig. 5.1 Number of new AI ethics principles by organization type, 2016–20. *Source* AI index report, 2021, p. 150, https://aipo-api.buddyweb.fr/app/uploads/2021/03/2021-AI-Index-Report.pdf#page=125&zoom=100,0,0

Justifiability of decisions and their contestability covers two areas—technical and legal. Developers should ensure the machines can justify their decisions, and lawyers should provide for a possibility to argue these decisions.[28] However, in human life, in addition to the absolutely right or wrong decisions, there are always those in-between. Cases of heavy moral decisions are numerous, and the people keep discussing them without finding the right solutions.[29] For example, it is generally expected that self-driving cars will reduce the number of accidents by strictly following the rules. However, in case of unforeseen accident and need to select between passengers and pedestrians on the road—which instruction will be encoded into the machine?

The number of ethical principles that the countries would like to see embedded in the new technologies is growing. The below graphs (Figs. 5.1 and 5.2) provide the visual representation of the number of the newly introduced ethical principles by countries and regions.

The organisations that have already published reports and guidelines on AI and ethics are numerous. Among them are the Institute of Electrical and Electronics Engineers (IEEE), the Organisation for Economic Co-operation and Development (OECD), the UN Secretary-General's High-Level Panel on Digital Cooperation, the International Telecommunications Union, the World Health Organization (WHO),[30]

---

[28] Khadem N. Tax office computer says yes, Federal Court says no. ABC. 8 October 2018.

[29] Awad E, Dsouza S, Kim R et al., The Moral Machine experiment, Nature, 2018, 563(7729): 59–64.

[30] World Health Organization, Ethics & Governance of Artificial Intelligence for Health, Guidance, 28 June 2021, https://www.who.int/publications/i/item/9789240029200.

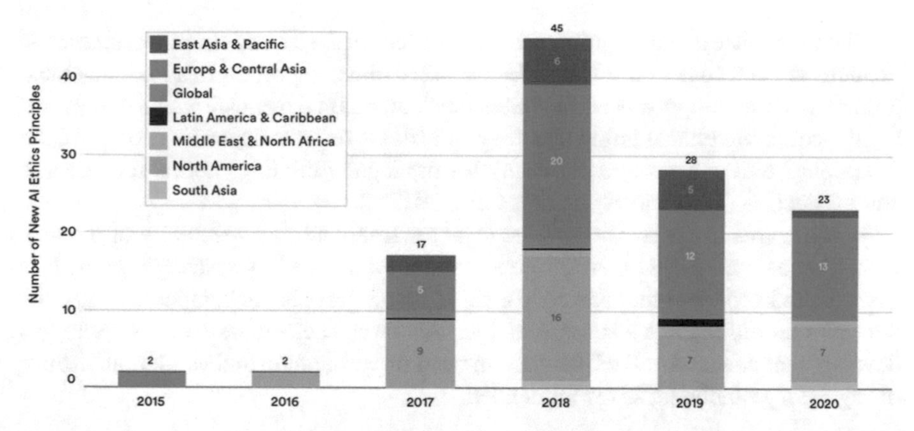

Fig. 5.2 Number of new ethics principles by region, 2015–20. *Source* AI index report, 2021, p. 150, at https://aipo-api.buddyweb.fr/app/uploads/2021/03/2021-AI-Index-Report.pdf#page=125&zoom=100,0,0

Council of Europe, European Union,[31] etc. UNESCO has initiated a global dialogue on the AI ethics. The Preliminary report on the first draft of the Recommendation on the Ethics of Artificial Intelligence was developed by a group of experts through broad consultations has been published on the UNESCO web-site for final discussions.[32]

Key guidance

Ensure that the development, deployment and use of AI systems meets the seven key requirements for Trustworthy AI: (1) human agency and oversight, (2) technical robustness and safety, (3) privacy and data governance, (4) transparency, (5) diversity, non-discrimination and fairness, (6) environmental and societal well-being and (7) accountability.[33]

Ethics Guidelines for Trustworthy AI, European Commission, 2019

[31] Report on artificial intelligence: questions of interpretation and application of international law in so far as the EU is affected in the areas of civil and military uses and of state authority outside the scope of criminal justice (2020/2013(INI)). Committee on Legal Affairs Rapporteur: Gilles Lebreton, 4.1.2021, p. 6, https://www.europarl.europa.eu/doceo/document/A-9-2021-0001_EN.pdf.

[32] Preliminary report on the first draft of the Recommendation on the Ethics of Artificial Intelligence, 2020, https://unesdoc.unesco.org/ark:/48223/pf0000374266.

[33] Ethics Guidelines for Trustworthy AI, Europen Commission, High-Level Expert Group on Artificial Intelligence, 8 April 2019, p. 2, https://digital-strategy.ec.europa.eu/en/library/ethics-guidelines-trustworthy-ai.

The abundance of ethical principles is criticised for the lack of clear measurements, benchmarks and consensus (The AI Index 2021 Annual report).[34] In addition, ethical guidelines do not imply any enforcement mechanisms and overestimation of the need for the corporate ethical guidelines may as well serve to "calm critical voices from the public" hiding a stronger necessity for the legally bindings norms for research and production (Hagendorff, 2020[35]; Calo, 2017[36]).

With regards to the revision of the legal environment, the specificity of the new technologies require that their progress cannot be blocked by restrictive laws. It is essential to keep the right balance to ensure the progress is not hampered. Thus, the new laws should be more flexible, scalable, holistic and adjustable to the technology's development needs. As all of them are in need of verification and validation, the use of regulatory sandboxes is recommended.

Google requests flexible laws and norms

Given the early stage of AI development, it is important to focus on **laws and norms that retain flexibility** as new possibilities and problems emerge. This is particularly crucial given that AI, like many technologies, is multi-purpose in nature.[37]

Google, Perspectives on issues in AI governance, 2019

Many of the international efforts are currently aimed towards development of globally acceptable frameworks and guidelines that take into consideration best practices of industries and governments, developing and introducing a governance system that would be flexible enough to consider the national legal systems and cultural differences, while ensuring respect of democratic values.

What world do we end up in?

Does AI usher in a new era of prosperity and international peace? Does it lead to shifts in the balance of power on the global stage, with attendant risks of conflict and miscalculation? Could AI lead to massive dislocation and a rise in political unrest, nationalism, and protectionism? Does AI concentrate power to control information in the hands of a few, or continue the democratization

[34] Daniel Zhang et al. "The AI Index 2021 Annual Report," AI Index Steering Committee, Human-Centered AI Institute, Stanford University, Stanford, CA, March 2021, at https://aipo-api.buddyweb.fr/app/uploads/2021/03/2021-AI-Index-Report.pdf.

[35] Hagendorff, T. (2020). The Ethics of AI Ethics: An Evaluation of Guidelines. Minds and Machines. https://doi.org/10.1007/s11023-020-09517-8.

[36] Calo Calo, R., Artificial intelligence policy: a primer and roadmap. SSRN Journal, 2017, 1–28..

[37] Google, Perspectives on issues in AI governance, 2019, p. 4, at https://ai.google/static/documents/perspectives-on-issues-in-ai-governance.pdf.

of information that computers, networks, and social media have unleashed? Does the cacophony of competing information lead to a turn away from truth to authoritarianism and tribalism, or does the wisdom of the crowds win out with a convergence on truth and centrist policies?

The technological opportunities enabled by artificial intelligence shape the future, but do not determine it. Nations, groups, and individuals have choices about how they employ and respond to various uses of AI. Their policy responses can guide, restrict, or encourage certain uses of AI.[38]

Michael C. Horowitz et al. (incl. P. Scharre), Artificial Intelligence and International Security, July 2018

Recognizing the omnipresent nature of the internet and cyberspace knowing no boundaries, the legal and ethical issues are broadly discussed at the international level, to become international standards, norms and laws. G20 Ministerial Statement on Trade and Digital Economy is appealing for "more innovation enabling approaches to policy making than in the past", for innovative-friendly governance, able to innovate itself.[39]

5.1.3 Debates Over Governance

Governance of AI-driven technologies is a complex issue of control and coordination covering the whole cycle of the development, utilization, recycling of AI-devices, control, capacity building and support to small and medium size enterprises, risk minimization, endorsement of regulatory frameworks and strategies, compliance with policies and procedures, and enforcement through regulatory agencies.[40] The governing process includes multiple stakeholders during the whole life-cycle of the AI products: research companies, developers, producers, distributors, interim agencies, governments. Governance also means organizing public awareness campaigns, conflict resolution, finding common interests, identifying priorities, setting timeframes for the integration of high technologies into our life through considering immediate and longer-term factors, developing road maps with steps and milestones. At the current stage it is challenging to ensure responsible governance as it is impossible to exercise control over the decisions taken in a black box mode.

[38] Michael C. Horowitz et al. (incl. P. Scharre), Artificial Intelligence and International Security, July 2018, Center for a New American Security's series on Artificial Intelligence and International Security, page 3 https://ethz.ch/content/dam/ethz/special-interest/gess/cis/center-for-securities-studies/resources/docs/CNAS_AI%20and%20International%20Security.pdf.

[39] G20 Ministerial Statement on Trade and Digital Economy, p. 4, https://www.meti.go.jp/press/2019/06/20190610010/20190610010-1.pdf.

[40] See materials of the research institutes: Future Society at Harvard Kennedy School, Future of Humanity Institute at the University of Oxford.

Governance: definitions

Governance has been defined to refer to structures and processes that are designed to ensure accountability, transparency, responsiveness, rule of law, stability, equity and inclusiveness, empowerment, and broad-based participation. Governance also represents the norms, values and rules of the game through which public affairs are managed in a manner that is transparent, participatory, inclusive and responsive. Governance therefore can be subtle and may not be easily observable. In a broad sense, governance is about the culture and institutional environment in which citizens and stakeholders interact among themselves and participate in public affairs.

International agencies such as UNDP, the World Bank, the OECD Development Assistance Committee (DAC) and others define governance as the exercise of authority or power in order to manage a country's economic, political and administrative affairs.

Good governance is expected to be participatory, transparent, accountable, effective and equitable and promotes rule of law....[41]

Among the common areas in the AI policies developed by governments,[42] there are **research and development** (national research centres, centres of excellence, investments), **education** (master's and doctoral programmes) **and jobs** (re-trainings), attracting talents (fees, visas), **industrial policies** (digital innovation hubs, support to start ups), **privacy**, **testing and evaluation** (regulatory sandboxes to test AI systems), **security** (policies, for infrastructure and data, including supply chains), **ethics** regulations (explainability, adaptation, tristfulness (digital infrastructure (publicly available datasets, tools and instructions in local languages), **standards** (international standards), **AI application in governments**, **international cooperation.**

Google urges to speed up development of international governance norms

While AI researchers, developers, and industry can lay the groundwork for what is technically feasible, it is ultimately up to government and civil society to determine the frameworks within which AI systems are developed and deployed. [...].

Given the open research culture in the AI field, increasing availability of functional building blocks (e.g., machine learning models for image recognition, speech-to-text, translation; processing hardware), and the usefulness of AI to many applications, AI technology is spreading rapidly. If the world

[41] Concept of Governance, UNESCO, http://www.ibe.unesco.org/en/geqaf/technical-notes/concept-governance.

[42] Strengthening international cooperation on artificial intelligence, Brookings, https://www.brookings.edu/research/strengthening-international-cooperation-on-artificial-intelligence/.

waits too long to establish international governance frameworks, **we are likely to end up with a global patchwork** that would slow the pace of AI development while also risking a race to the bottom. A self-regulatory or co-regulatory set of international governance norms that could be applied flexibly and adaptively would enable policy safeguards while preserving the space for continued beneficial innovation.[43]

Google, Perspectives on issues in AI governance, 2019

New and promising technologies always attract government and commercial investments, whether it is for profit, military superiority, or to improve quality of life for the nation. As soon as artificial intelligence started showing potential, the funding immediately flooded this futuristic technology.

AI victory in Go inspired investments in new technologies

AlphaGo, the AI system developed by Google-owned DeepMind.

In 2016, AlphaGo beat South Korean master Lee Se-dol at the ancient Chinese board game Go, and in May this year, it toppled the Chinese world champion, Ke Jie. Two professors who consult with the Chinese government on AI policy told *The New York Times* that these games galvanized the country's politicians to invest in the technology. And the report makes China's ambitions in this area clear: the country says it will become the world's leader in AI by 2030.[44]

The Organization for Economic Cooperation and Development (OECD) AI Policy Observatory collects information on the national AI initiatives presenting them on its website. As of the time of this book writing, OECD repository contains 600 AI policy initiatives from 60 countries. They cover the issues of investments, support to innovations, education of experts and re-education of those losing jobs, norms, standards, policies, procedures. In its turn, OECD conducts research and promotes the value-based principles for the new technology approved by its 38 member countries, that include "inclusive growth, sustainable development and wellbeing, human centred values and fairness, transparency and explainability, robustness, security and safety, accountability".

[43] Google, Perspectives on issues in AI governance, 2019, p. 4. https://ai.google/static/documents/perspectives-on-issues-in-ai-governance.pdf.

[44] James Vincent, China and US are battling to become the world first AI superpower, 3 August 2017, *The Verge*, https://www.theverge.com/2017/8/3/16007736/china-us-ai-artificial-intelligence.

Table 5.1 Funding of selected AI strategies as per 2018

Country/Region	Release date	Official strategy	Funding in local currency (July 2018 USD exchange rates)
Australia	May 2018	Australian technology and science growth plan	29.9 million (21.6 million USD)
Canada	March 2017	Pan-Canadian artificial intelligence strategy	125 million (95 million USD)
Singapore	May 2017	AI Singapore	150 million over 5 years (91.5 million USD)
Denmark	January 2018	Strategy for Denmark's digital growth	75 million in 2018, 125 million each year to 2025 (11.7 million USD, 19.5 million USD)
Taiwan	January 2018	Taiwan AI action plan	36 billion over four years (1.18 billion USD)
France	March 2018	France's strategy for AI	1.5 billion over five years (1.75 billion USD)
EU Commission	April 2018	Communication artificial intelligence for Europe	Increase annual investment in AI to 1.5 billion by end of 2020 (1.75 billion USD)
United Kingdom	April 2018	Industrial strategy: artificial intelligence sector deal	950 million from government, academia, and industry (1.24 billion USD)
South Korea	May 2018	Artificial intelligence R&D strategy	2 trillion (1.95 billion USD)

Source The 2020 CIFAR Report on National and Regional AI Strategies[47]

In 2017, Canada became the first country to introduce the AI Strategy. In 2020, the OECD AI Policy Observatory reported the number of countries committed to developing AI strategies (government policies) increased to 30, some of them with a multi-billion-dollar budgets.[45] OECD also provides recommendations for policy makers, such as "investing in AI research and development, fostering a digital ecosystem for AI, providing an enabling policy environment for a AI, building human capacity and preparing for labour market transition, international corporation for trustworthy AI".[46]

Table 5.1 presents the AI strategies of nine countries. The budgets vary greatly, e.g., Australia's budget is 25 million USD, while the one of South Korea is 2 billion USD.

[45] OECD AI Policy Observatory, The 2021 AI Index: AI Industrialization and Future Challenges, https://www.oecd.ai/wonk/2021-ai-index-highlights.

[46] OECD AI Principles overview, OECD AI Policy Observatory, https://www.oecd.ai/ai-principles.

[47] Tim Dutton, Building an AI World: Report on National and Regional AI Strategies, CIFAR, 2020, p. 5, at https://cifar.ca/wp-content/uploads/2020/05/buildinganaiworld_eng.pdf.

Table 5.2 AI strategies heat map

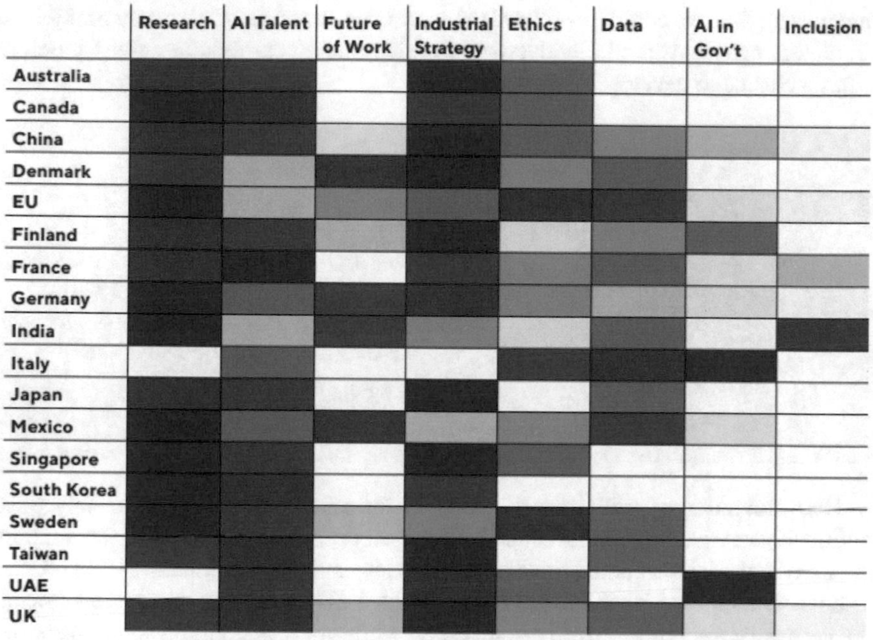

	Research	AI Talent	Future of Work	Industrial Strategy	Ethics	Data	AI in Gov't	Inclusion
Australia								
Canada								
China								
Denmark								
EU								
Finland								
France								
Germany								
India								
Italy								
Japan								
Mexico								
Singapore								
South Korea								
Sweden								
Taiwan								
UAE								
UK								

Source The 2020 CIFAR Report on National and Regional AI Strategies[48]

The 2020 CIFAR Report concludes that the governments strategies towards integrating AI technologies vary, and each of them is unique. Table 5.2 maps the areas that covered by the 18 strategies (developed and under development by the time of data collection) as important for accommodating the AI technologies needs and the level of attention to them (the darker is the colour the more attention and budgeting this area receives).

As demonstrated by Table 5.2, governments give priority to industrialization, with the research being on the second place. AI in Governments received less attention, and inclusion had the lowest coverage (only India and France). This research allows to conclude that the efforts of international organizations have to be aimed at ensuring harmonization of the countries' policies, to avoid any kind of gaps.

Currently, there is an increasing trend for creating advisory boards of technical experts (who can be international experts), task forces and oversight committees providing their advice to politicians, non-technical experts taking decisions. For example, Israel, Singapore, Canada created and strengthened connections between governments, academia, businesses, thus boosting their AI potential.[49] AI implementation in the **public sector** requires taking informed decisions by politicians, based on

[48] Tim Dutton, Building an AI World: Report on National and Regional AI Strategies, CIFAR, 2020, p. 5, at https://cifar.ca/wp-content/uploads/2020/05/buildinganaiworld_eng.pdf.

[49] OECD AI policy Observatory, https://www.oecd.ai/.

strategic vision, short and longer-term plans. The speed of its implementation is lower and weaker than we could have[50] and there are even concerns that "overly prohibitive regulatory proposals could inadvertently undermine this promising industry before it has a chance to develop."[51]

AI: to use or not to use

In setting benchmarks, it is important to factor in the opportunity cost of not using an AI solution when one is available; and to determine at what levels of relative safety performance AI solutions should be used to supplement or replace existing human ones. AI systems can make mistakes, but so do people, and in some contexts AI may be safer than alternatives without AI, even if it is not fail-proof.[52]

Google, 2019

High technologies, with their potential for influencing the whole of society, pose multiple issues that the governments alone may not be effective to resolve. This entails the need to the whole society involvement, engagement of communities and NGOs. The newly identified trends in having a so-called **networked governance** allowing the necessary level of **flexibility in learning** while going and taking solutions based on **peer reviews,** and not on the hierarchical command, matches perfectly well the needs for the integration of the flexible, adaptive and scalable technologies.

There are also strong messages in support of a more courageous research: "experimentation with new technologies and business models should generally be permitted by default. Unless a compelling case can be made that a new invention will bring serious harm to society, innovation should be allowed to continue unabated, and problems, if they develop at all, can be addressed later."[53] Good governance in the era of high technologies also means that the technological progress is not limited.

Successful governance of the new technologies is equally responsible for minimizing risks through introducing specific mechanisms and regulations agreed globally. With full understanding that it is not easy to map risks against for example ethical principles, one of the options may be the development of a range of risks

[50] Cath C, Wachter S, Mittelstadt B, Taddeo M, Floridi L. Artificial Intelligence and the 'Good Society': the US, EU, and UK approach. Sci Eng Ethics. 2018 Apr;24(2):505–528. https://doi.org/10.1007/s11948-017-9901-7. Epub 2017 Mar 28. PMID: 28,353,045.

[51] Adam Thierer, Andrea Castillo O'Sullivan, and Raymond Russell, Artificial Intelligence and Public Policy, Mercatus Research, George Mason University, 2017, https://doi.org/10.13140/RG.2.2.14942.33604.

[52] Google, Perspectives on issues in AI governance, Google, 2019, p. 19, https://ai.google/static/documents/perspectives-on-issues-in-ai-governance.pdf.

[53] Adam Thierer, Permissionless Innovation: The Continuing Case for Comprehensive Technological Freedom, 2nd ed. (Arlington, VA: Mercatus Center at George Mason University, 2016).

from minimal to critical against every agreed ethical principle with an explanation of consequences.

The 2019 Doha debates: voting results against key statements

The 2019 Doha debates with participation of Nick Bostrom, demonstrated that people would like to see more informed politicians and a stronger oversight over AI. The voting results were the following.

1. AI will create more equality among nations: 12.83%
2. Politicians need to understand AI: 33.75%
3. AI could destroy humanity as we know it 18.11%
4. Without oversight, AI will amplify inequality 35.31%

Artificial Intelligence, Full debate, Doha Debates,[54] *3 April 2019*

Recognizing that countries have different levels of development, there is a risk of technological gaps that may further increase inequalities. Production and accommodation of high technologies is very expensive. It includes expensive powerful equipment, high salaries of AI specialists and security experts; need for the best quality data suiting specific purposes. This issue should be resolved jointly through multisectoral discussions, national plans of action, their costing and funding. International development organizations could be also engaged in providing support with resource mobilization, expertise and partnerships.

All the above-mentioned factors are equally relevant for the AI technologies used for protecting national sovereignty and interests. However, one factor remains specific. Militarized AI can take life and death decisions, and all the enumerated risks and threats multiply enormously.

5.1.4 Debates Over Military Use of AI Offensive Capabilities

Competing for domination in this world, but also to protect the national interests, the states are accumulating AI-empowered weapons. Continuous competition in applying innovative technology for military purposes leads to an arms race, resulting in further escalation of international tensions and conflicts. Being driven by the national security interests, the researchers abstain from sharing critical findings, or contributing to discussions of AI control.[55]

---

[54] Artificial Intelligence, Full debate, Doha Debates, 3 April 2019, 47:15, https://www.youtube.com/watch?v=ffG2EdYxoCQ.

[55] Bostrom [2]; Armstrong, Bostrom, and Shulman, Racing to the Precipice, 2016.

With the skyrocketing technological developments, both governments and private sector increase investments into AI research competing for leadership and profits. The major areas of military investments are data collection and data mining; identification and tracking of potential threats; autonomous vehicles for defence and military operations; autonomous vehicles for patrol, scouting, logistics, and payload delivery; human enhancement, AI for augmentation of capabilities of human operators.

Among the latest notorious projects for military and surveillance are the DARPA AI projects on Military Imaging and Surveillance Technology-Long Range (MIST-LR) Program (2013—"to develop a fundamentally new optical Intelligence, Surveillance, and Reconnaissance (ISR) capability able to provide high-resolution 3-D images to locate and identify a target",[56] "to build sensors for aircraft and ground vehicles that will allow for target characterization beyond the physical-aperture diffraction-limit of the receiver system,… focusing on new electro-optical sensing methods based on computational imaging; synthetic-aperture imaging; digital holography."[57]

The 2020 RAND Report on Maintaining the Competitive Advantage in Artificial Intelligence and Machine Learning recognized that currently it is impossible to name a leading country in the AI development. Instead, it proposes to discuss the leading role in some specific areas. Comparing the US and China approaches, it stresses that the US is leading in some key areas, while China demonstrates a much stronger political will.

Competitive comparison in AI and Machine Learning: China and the United States

- As of early 2020, the United States has a modest lead in AI technology development because of its substantial advantage in the advanced semiconductor sector. China is attempting to erode this edge through massive government investment. The lack of a substantial U.S. industrial policy also works to Chinese advantage.
- China has an advantage over the United States in the area of big data sets that are essential to the development of AI applications. This is partly because data collection by the Chinese government and large Chinese tech companies is not constrained by privacy laws and protections. However, the Chinese advantage in data volume is probably insufficient to overcome the U.S. edge in semiconductors.

---

[56] https://www.darpa.mil/program/military-imaging-and-surveillance-technology#:~:text=The%20Military%20Imaging%20and%20Surveillance,possible%20with%20existing%20optical%20systems.

[57] https://www.militaryaerospace.com/communications/article/16715144/darpa-moves-to-phasetwo-of-mistlr-longrange-eo-imaging-technology-project.

- Breakthrough fundamental research is not a critical dimension for comparing U.S.-China relative competitive standing from a DoD perspective. Fundamental research, regardless of whether it is U.S., Chinese, or a U.S.-Chinese collaboration, is available to all.
- Commercial industry is also not a critical dimension for competitive comparison. Industries with corporate headquarters in the United States and in China seek to provide products and services wherever the market is.[58]

RAND Report, Maintaining the Competitive Advantage in Artificial Intelligence and Machine Learning, 2020

Development and implementation of the Autonomous Weapon Systems (AWSs) have entailed heated discussions world-wide. Technical experts and engineers are aware of the immaturity of the new AI-enabled technologies, their vulnerability to cyber attacks, complexity in defining autonomy at the modern level of computers' development. Legal experts recognize the lack of norms and regulations with specific definitions, responsibility levels, including under International Law, clear definition of the levels of human control and mode of use in various operational scenarios. Ethicists are raising loud concerns around the issue of algorithms taking life and death decisions and dehumanizing dignity; media reports keep alerting the world on the application of the AI-empowered weapons and civil society activists are appealing for urgency of action.

Positive obligation of Third Parties

Common Article 1 of the Geneva Conventions requires that states 'respect and ensure respect for' the Geneva Conventions 'in all circumstances'. In the new 2016 Commentary to the Convention, the existence of not only a negative obligation, but also a positive obligation of third countries to a conflict to prevent violations was confirmed. Hence, third countries must do everything 'reasonably in their power to prevent and bring such violations to an end'. The use of autonomous weapons systems (AWS) is imminent in the future, as demonstrated by the Pentagon committing to spend $2 billion on research, with similar research programmes taking place in other countries. The buying and selling of these AWS is an equally impending part of the future. Consequently, inevitably a state that is buying or being supplied with AWS will use them in a conflict. Therefore, suppliers of such systems will have to comply with the aforementioned positive obligation. [...].

[58] Waltzman, R., Ablon, L., Curriden, C., Hartnett, G. S., Holliday, M. A., Ma, L., Nichiporuk, B., Scobell, A. & Tarraf, D. C. (2020) Maintaining the Competitive Advantage in Artificial Intelligence and Machine Learning, 2020, at https://www.rand.org/pubs/research_reports/RRA200-1.html.

One of the most critical questions is "whether it will be their [suppliers] responsibility at the manufacturing stage to ensure that the system cannot violate the Geneva Conventions and—because autonomous systems are somewhat uncontrollable and unpredictable as they will also learn rather than only carrying out pre-programmed commands—whether the supplying state will be obligated to maintain a permanent tether to the supplied AWS to monitor them. The implications of tethering the supplied AWS may go well beyond ensuring compliance with international humanitarian law (IHL), and may include multiplying the leverage of the supplying state by turning the systems into 'cyber mercenaries'.[59]

Counterarguments come from the military, who recognize the AWSs strategic and operational advantages, cost-efficiency, as well as their role in protecting the life of soldiers in military operations. The advantages include a high speed of implementation, with no time loss for remote controls or communication disruptions, swarming capacities, force multiplication, etc.

Questions around AWS

[T] he issue surrounding autonomous weapons is, what happens when a predator drone has as much autonomy as a self-driving car? What happens when you have a weapon that's out in the battlefield, and it's making its own decisions about whom to kill? Is that something that we're comfortable with? What are the legal and moral and ethical ramifications of this? And the strategic implications? What might they do for the balance of power between nations, or stability among countries? These are really the issues surrounding autonomous weapons, and it's really about this idea that we might have, at some point of time and perhaps the not very distant future, machines making their own decisions about whom to kill on the battlefield.[60]

Paul Scharre, Interview to the Future of Life Institute, 2020

The most heated discussions have been and are still ongoing around the **Lethal Autonomous Weapon Systems** (LAWS), that are "a special class of weapon systems that use sensor suites and computer algorithms to independently identify a target and employ an onboard weapon system to engage and destroy the target without

[59] A. Kajander, A. Kasper and E. Tsybulenko, "Making the Cyber Mercenary—Autonomous Weapons Systems and Common Article 1 of the Geneva Conventions," *2020 12th International Conference on Cyber Conflict (CyCon)*, 2020, pp. 79–95, https://doi.org/10.23919/CyCon49761.2020.9131722.

[60] Paul Scharre, Interview to the Future of Life Institute, March 16, 2020, (by Lucas Perry). https://futureoflife.org/2020/03/16/on-lethal-autonomous-weapons-with-paul-scharre/#:~:text=Paul%20Scharre%3A%20Yes%2C%20so%20autonomy,for%20some%20period%20of%20time.

manual human control of the system."[61] The issue of LAWS is debated at the level of technical experts, highest national and international political levels, by regional and global intergovernmental organizations seeking for consensus in developing an international framework on autonomous weapon systems.[62] There is a number of letters and petitions, addressed to the higher internationals authorities, to the United Nations, aiming to ban LAWS as the AI cannot be entrusted with taking a decision over human life.[63]

> [M]achines with the power and discretion to take lives without human involvement are politically unacceptable, morally repugnant and should be prohibited by international law.[64]
>
> *UN Secretary-General António Guterres, 25 March 2019*

The delineation between the right and wrong in the use of AWSs includes the relevance to the International Humanitarian Law.[65] The global opinions are split on whether the application of autonomous weapons should be under the Convention on Certain Conventional Weapons or under any other convention that is yet to be developed. The internationally applicable principles codified in the Geneva and the Hague conventions, include the principle of distinction protecting civilians; principle of proportionality protecting civilian infrastructure; principle of avoiding unnecessary suffering of combatants protecting wounded and surrendered, etc. In addition, there exist uncodified but applied ethical, moral principles and pure human feelings respecting the right to life as a highest value that equally play their role in any war.[66] The issue of the war dehumanization, fundamental issues of human dignity as enshrined in the Universal Declaration of Human Rights (see its Preamble)

---

[61] Defense Primer: U.S. Policy on Lethal Autonomous Weapon Systems, Congressional research Service, 1 December 2020, at https://fas.org/sgp/crs/natsec/IF11150.pdf.

[62] Mary Wareham, "As Killer Robots Loom, Demands Grow to Keep Humans in Control of Use of Force", Human Rights Watch, 2020, available at: www.hrw.org/world-report/2020/country-chapters/killerrobots-loom-in-2020.

[63] Future of Life Institute, "Autonomous Weapons: An Open Letter from AI and Robotics Researchers", 28 July 2015, at https://futureoflife.org/open-letter-autonomous-weapons/; "An Open Letter to the United Nations Convention on Certain Conventional Weapons", 21 August 2017 at: https://futureoflife.org/autonomous-weapons-open-letter-2017/.

[64] Machines Capable of Taking Lives without Human Involvement Are Unacceptable, Secretary-General Tells Experts on Autonomous Weapons Systems, SG/SM/19512-DC/3797, 25 March 2019. https://www.un.org/press/en/2019/sgsm19512.doc.htm.

[65] United States, Implementing International Humanitarian Law in the Use of Autonomy in Weapon Systems: Working Paper Submitted by the United States of America, UN Doc. CCW/GGE.1/2019/WP.5, 28 March 2019, available at: https://tinyurl.com/y4xe7tmc.

[66] Elvira Rosert and Frank Sauer, "Prohibiting Autonomous Weapons: Put Human Dignity First", Global Policy, Vol. 10, No. 3, 2019.

and respect to it have been vastly discussed by the expert community,[67,68,69] civil society,[70] ICRC,[71] etc.

Israel's operation against Hamas was the world's first AI war

"Having relied heavily on machine learning, the Israeli military is calling Operation Guardian of the Walls the first artificial-intelligence war."

For the first time, artificial intelligence was a key component and power multiplier in fighting the enemy," an IDF Intelligence Corps senior officer said. "This is a first-of-its-kind campaign for the IDF. We implemented new methods of operation and used technological developments that were a force multiplier for the entire IDF."

[..]

The mapping of Hamas's underground network was done by a massive intelligence-gathering process that was helped by the technological developments and use of Big Data to fuse all the intelligence. Once mapped, the IDF was able to have a full picture of the network both above and below ground with details, such as the depth of the tunnels, their thickness and the nature of the routes. With that, the military was able to construct an attack plan that was used during the operation.[72]

The Jerusalem Post, 27 May 2021

The Human Rights Council raised the issue of ethical implications and human dignity violations in 2013.[73] The intergovernmental discussions of LAWS were initiated in 2014 within the framework of the United Nations (UN) Convention on Certain Conventional Weapons (CCW) that is aimed "to ban or restrict the use of specific types of weapons that are considered to cause unnecessary or unjustifiable suffering

---

[67] Deane-Peter Baker, "The Awkwardness of the Dignity Objection to Autonomous Weapons", The Strategy Bridge, 6 December 2018, at: https://thestrategybridge.org/the-bridge/2018/12/6/theawkwardness-of-the-dignity-objection-to-autonomous-weapon.

[68] Amanda Sharkey, "Autonomous Weapons Systems, Killer Robots and Human Dignity", Ethics and Information Technology, Vol. 21, No. 2, 2019.

[69] Christof Heyns, "Autonomous Weapons in Armed Conflict and the Right to a Dignified Life: An African Perspective", South African Journal on Human Rights, Vol. 33, No. 1, 2017, pp. 62–63.

[70] KRC, Making the Case: The Dangers of Killer Robots and the Need for a Preemptive Ban, 2016, pp. 21–25.

[71] ICRC, Ethics and Autonomous Weapon Systems: An Ethical Basis for Human Control?, Geneva, 3 April 2018, available at: www.icrc.org/en/download/file/69961/icrc_ethics_and_autonomous_weapon_systems_report_3_april_2018.pdf.

[72] Anna Ahronheim, Israel's operation against Hamas was the world's first AI war, May, 2021, The Jerusalem Post, at https://www.jpost.com/arab-israeli-conflict/gaza-news/guardian-of-the-walls-the-first-ai-war-669371.

[73] Christof Heyns, Report of the Special Rapporteur on Extrajudicial, Summary or Arbitrary Executions, UN Doc. A/HRC/23/47, 2013, p. 17, available at: https://digitallibrary.un.org/record/755741/files/A_HRC_23_47-EN.pdf.

to combatants or to affect civilians indiscriminately".[74] The talks had initially an informal character, and since 2017 the official Group of Governmental Experts (GGE) started its work that is now aimed at developing "aspects of the normative and operational framework" on LAWS.[75] After a number of rounds and reports, in 2021, there is yet no common international agreement among the CCW States Parties. For more detailed information on the GGE work, see our previous book *Cyber Arms: Security in Cyber Space*.[76]

Among the issues that have been discussed are the development of commonly agreed definitions of autonomy, autonomous functions and their most critical stages, separation of debates of LAWS from discussing other types of autonomous weapons (e.g., drones), standards for the LAWS categorization and classification, and their design, etc.[77] It has been recognized that the weapon systems are no longer a set of hardware devices ranked under certain standards, that their AI capacities and hardware are in constant evolution and changes, their constituting parts and control systems may be separated and be in different locations.[78]

The definitions of LAWS vary, but mostly refer to the role of the human operator in their critical functionality. The EU Report on artificial intelligence defines them as "weapons systems without meaningful human control over the critical functions of targeting and attacking individual targets.[79] Two elements were identified as a common ground for discussions by the majority of the CCW States Parties—critical functions in autonomy and meaningful human control in them.[80] However, it is equally recognized that the presence of the human control impacts the weapons operational capacities, decreasing the speed of reaction, and this may be objected by military.

---

[74] United Nations, The Convention on Certain Conventional Weapons, https://www.un.org/disarmament/the-convention-on-certain-conventional-weapons/.

[75] UN, Meeting of the High Contracting Parties to the Convention on Prohibitions or Restrictions on the Use of Certain Conventional Weapons Which May Be Deemed to Be Excessively Injurious or to Have Indiscriminate Effects: Revised Draft Final Report, UN Doc. CCW/MSP/019/CRP./Rev.1, Geneva, 15 November 019 (CCW Meeting Final Report), p. 5.

[76] Chapter 2.7 Cyber Arms Race and Control, pp. 141–179, in S. Abaimov, M. Martellini "Cyber Arms And Security In Cyber Space", 2020.

[77] UN, Report of the 2019 Session of the Group of Governmental Experts on Emerging Technologies in the Area of Lethal Autonomous Weapons Systems: Chair's Summary, UN Doc. CCW/GGE.1/2019/3/Add.1, 8 November 2019, p. 3, at: https://documents.unoda.org/wp-content/uploads/2020/09/CCW_GGE.1_2019_3_E.pdf.

[78] ICRC, Autonomous Weapon Systems: Implications of Increasing Autonomy in the Critical Functions of Weapons, Geneva, 2016; US Department of Defense (DoD), Directive 3000.09, "Autonomy in Weapon Systems", 2012 (amended 2017).

[79] Report on artificial intelligence: questions of interpretation and application of international law in so far as the EU is affected in the areas of civil and military uses and of state authority outside the scope of criminal justice (2020/2013(INI)). Committee on Legal Affairs Rapporteur: Gilles Lebreton, 4.1.2021, p. 11, https://www.europarl.europa.eu/doceo/document/A-9-2021-0001_EN.pdf.

[80] Richard Moyes, "Key Elements of Meaningful Human Control", Article 36, April 2016, available at: www.article36.org/wp-content/uploads/2016/04/MHC-2016-FINAL.pdf. Article 36 is a member of the KRC.

Agreement on common approaches to weapon systems based on the functionalist approach and human control[81] and responsibility is already a meaningful step forward (see CCW principle "[h]uman responsibility for decisions on the use of weapons systems must be retained since accountability cannot be transferred to machines",[82] SIPRI and ICRC report[83], 2020). The following step would be to agree on the **levels and types of this control**, and weapons s**tandardization**. Considering the impossibility of developing standards embracing all AWS, a more individual approach is needed built on case-by-case studies to develop conceptual approaches to the human control implementation (scenarios).

The academic community has provided a vast contribution to the discussions, researching in the areas of AI potential and vulnerabilities, digital autonomy, AI-empowered weapons and their influence on the international stability and war.

The literature review shows that the growing number of autonomous weapons, in addition to strategic advantages, creates another level of threats related to vulnerabilities of AI-systems. It is imperative, and in the interests of the great powers as well, that the decision be taken on their regulation .[84]

European Union has a strong position towards LAWS, recognizing that they must fall under the provisions of the Convention on Certain Conventional Weapons. Its Resolution of 12 September 2018 requires that "attacks should always be carried out with significant human intervention".[85] The EU Report on artificial intelligence "[c]onsiders that the use of lethal autonomous weapon systems raises fundamental ethical and legal questions about the ability of humans to control these systems, and requires that AI-based technology should not be able to make autonomous decisions involving the legal principles of **distinction, proportionality and precaution**;"[86] It urges its Member States to come up with a joint "common position on autonomous weapons systems that ensures meaningful human control over the critical functions of weapons systems" and *act accordingly*. This report equally requests the EU Member

[81] Ilse Verdiesen, Filippo Santoni de Sio and Virginia Dignum, "Accountability and Control over Autonomous Weapon Systems: A Framework for Comprehensive Human Oversight", Minds and Machines, 2020, available at: https://link.springer.com/article/10.1007/s11023-020-09,532-9.

[82] CCW Meeting Final Report, 2019.

[83] Vincent Boulanin et al., Limits on Autonomy in Weapon Systems: Identifying Practical Elements of Human Control, June 2020, SIPRI and ICRC, https://www.sipri.org/sites/default/files/2020-06/2006_limits_of_autonomy.pdf.

[84] For the general argument, see Hedley Bull, The Anarchical Society: A Study of Order in World Politics, Macmillan, London, 1977. For the case of AI, see Elsa B. Kania and Andrew Imbrie, "Great Powers Must Talk to Each Other about AI", Defense One, 28 January 2020, available at: www.defenseone.com/ideas/2020/01/great-powers-must-talk-each-other-about-ai/162686/?oref=d-river.

[85] Autonomous weapon systems, European Parliament resolution of 12 September 2018 on autonomous weapon systems, (2018/2752(RSP)), http://www.europarl.europa.eu/sides/getDoc.do?pubRef=-//EP//NONSGML+TA+P8-TA-2018-0341+0+DOC+PDF+V0//EN.

[86] Report on artificial intelligence: questions of interpretation and application of international law in so far as the EU is affected in the areas of civil and military uses and of state authority outside the scope of criminal justice (2020/2013(INI)). Committee on Legal Affairs Rapporteur: Gilles Lebreton, 4.1.2021, p. 11, https://www.europarl.europa.eu/doceo/document/A-9-2021-0001_EN.pdf.

States to conduct a joint assessment of AWSs to understand the benefits versus threats to these systems from cyber attacks or malfunctions. Considering their high risk, the AWS should be allowed only under a special authorization for *clearly defined cases*. Anthropomorphizing of weapons should be also prohibited.

EU Report on Artificial Intelligence

Stresses the importance [...] of European Union involvement in the creation of an international legal framework for the use of artificial intelligence: urges the EU to take the lead and assume, with the United Nations and the international community, an active role in promoting this global framework governing the use of AI for military and other purposes, ensuring that this use remains within the strict limits set by international law and international humanitarian law, in particular the Geneva Conventions of 12 August 1949; [...].

Calls on the Vice President of the Commission / High Representative for Foreign Affairs and Security Policy to pave the way for global negotiations with a view to putting in place an AI arms control regime and updating all existing treaty instruments on arms control, disarmament and non-proliferation so as to take into account AI-enabled systems used in warfare; [...].

Points to the clear risks involved in decisions made by humans if they rely solely on the data, profiles and recommendations generated by machines; points out that the overall design of AI systems should also include guidelines on human supervision and oversight; calls for an obligation to be imposed regarding transparency and explainability of AI applications and the necessity of human intervention, as well as other measures, such as independent audits and specific stress tests to facilitate and enforce compliance; [...].

Emphasises the importance of verifying how high-risk AI technologies arrive at decisions; recalls that the principles of non-discrimination and proportionality need to be respected, and that questions of causality, liability and responsibility, as well as transparency, accountability and explainability, need to be clarified to determine whether, or to what extent, the state as an actor in public international law, but also in exercising its own authority, can act with the help of AI-based systems with a certain autonomy, without breaching obligations stemming from international law, such as due process [...].

Insists that AI systems must always comply with the principles of responsibility, equity, governability, precaution, accountability, attributability, predictability, traceability, reliability, trustworthiness, transparency, explainability, the ability to detect possible changes in circumstances and operational environment, the distinction between combatants and non-combatants, and proportionality; stresses that the latter principle makes the legality of a military action conditional on a balance between the objective pursued and the means used, and that the assessment of proportionality must always be made by a human being.[87]

AI experts express hope that in the far future and under certain conditions smart weapons will definitely be able to distinguish between military and civilians and may potentially be better in reducing unnecessary damages. But this is still a far future. In connection with the above discussion, it is worth reminding that the level of responsibility of researchers developing and applying various algorithms should be also augmented. They should be able to raise their concerns without being scared to lose their jobs. Internationally supported protection measures should be developed for this specific case.

The role of academia, teachers in education the computer specialists, experts in AI-development and related fields is crucial in raising concerns while educating future generations.

> … [T]he computer is a powerful new metaphor for helping us to understand many aspects of the world, … it enslaves the mind that has no other metaphors and few other resources to call on. The world is many things, and no single framework is large enough to contain them all, neither that of man's science nor that of his poetry, neither that of calculating reason nor that of pure intuition. And just as a love of music does not suffice to enable one to play the violin—one must also master the craft of the instrument and of music itself—so is it not enough to love humanity in order to help it survive. The teacher's calling to teach his craft is therefore an honourable one. But he must do more than that: he must teach more than one metaphor, and he must teach more by the example of his conduct then by what he writes on the blackboard. He must teach the limitations of his tools as well as their power.[88]
>
> *Joseph Weizenbaum, 1976*

This equally refers to decisions taken by developers and producers.

---

[87] Report on artificial intelligence: questions of interpretation and application of international law in so far as the EU is affected in the areas of civil and military uses and of state authority outside the scope of criminal justice (2020/2013(INI)). Committee on Legal Affairs, Rapporteur Gilles Lebreton, 4.1.2021, pp. 9–10, https://www.europarl.europa.eu/doceo/document/A-9-2021-0001_EN.pdf.

[88] Weizenbaum [3], p. 277.

Project Maven precedent

A project to use machine learning to identify people and objects from UAV videos was tasked to Google by the Department of Defense. A soon as this information became available to Google employees, they refused to continue the work and the project was not renewed.

Similar case took place in Microsoft.

To conclude this review, it is worth highlighting that the international regulations on the autonomous weapon systems is a necessary condition towards mitigating risks. The development and spread of AI-enabled autonomous functions will continue evolving. From the cyber-security perspective, the following can be recommended as the basis of these regulations.

- Human supervision should always be maintained until the technology is deemed sufficiently trusted to be fully autonomous. It is essential in autonomous weapons in the most critical functions. Understanding that there may always be exceptional situations that may need the approval of the respective government agency, with full understanding of risks.
- Development of internationally recognized standards of safety testing, evaluation, verification and validation on a case by case basis or within the established frameworks based on control schemes. (e.g., communication protocols for drones, AWS network architecture).

The 2020 RAND report on Maintaining the Competitive Advantage in Artificial Intelligence and Machine Learning recommends to:

Create and tailor verification, validation, and evaluation techniques for AI technologies.

Create development, test, and evaluation processes for new operational concepts that employ AI.[89]

- Full autonomy is recommended in exceptional cases, e.g., in disaster relief, for hazardous operations without possibility of remote control, defence from missiles attacks, in in peacekeeping missions, simulation exercises, areas with poor signal strength or coverage (i.e., desserts, deep ocean, space). High risks of unidentified malware insider the system should be recognized, as it can lead to unexpected consequences. The use of these fully autonomous systems should be authorized by governments.

[89] Waltzman, R., Ablon, L., Curriden, C., Hartnett, G. S., Holliday, M. A., Ma, L., Nichiporuk, B., Scobell, A. & Tarraf, D. C. (2020) Maintaining the Competitive Advantage in Artificial Intelligence and Machine Learning, 2020, at https://www.rand.org/pubs/research_reports/RRA200-1.html.

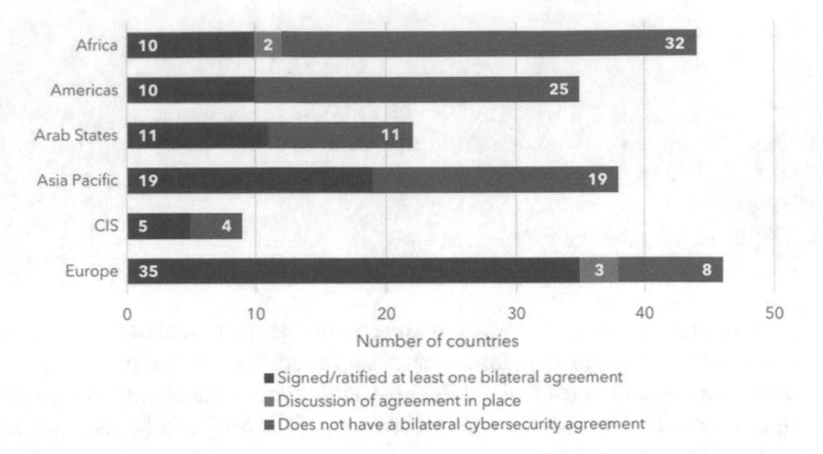

Fig. 5.3 Countries participating in bilateral cybersecurity agreements. *Source* Global Cybersecurity Index, ITU, 2021, p. 20

The need for the **new approaches** in creating the legal framework for the emerging technologies is recognized.[90] Technical experts should propose innovative evaluation solutions to ensure high quality of verification and validation, and political will is a decisive factor in developing and ratifying the international framework on the AI-empowered weapons.

The encouraging factor is that humanity has already gained experience in developing regulations on chemical, nuclear and biological weapons. With the new technologies, it is time to do the same for cyber weapons.

5.2 Multilateral Collaboration for Peaceful AI

The new technologies are a world-wide phenomenon and the cyberspace does not know geographical borders. To ensure the necessary level of protection from cyber threats countries conclude bilateral and multilateral agreements. The ITU annual report on the Global cybersecurity presents the overview of the countries that adopted bilateral and multilateral agreements in 2020 (Figs. 5.3 and 5.4).

As per the analysed data, the bilateral agreements have been concluded between 90 countries. The agreements content differs and vary from capacity development, to legal assistance and sharing information.

[90] Elvira Rosert and Frank Sauer, "How (Not) to Stop the Killer Robots: A Comparative Analysis of Humanitarian Disarmament Campaign Strategies", Contemporary Security Policy, 30 May 2020, available at: https://www.tandfonline.com/doi/full/10.1080/13523260.2020.1771508.

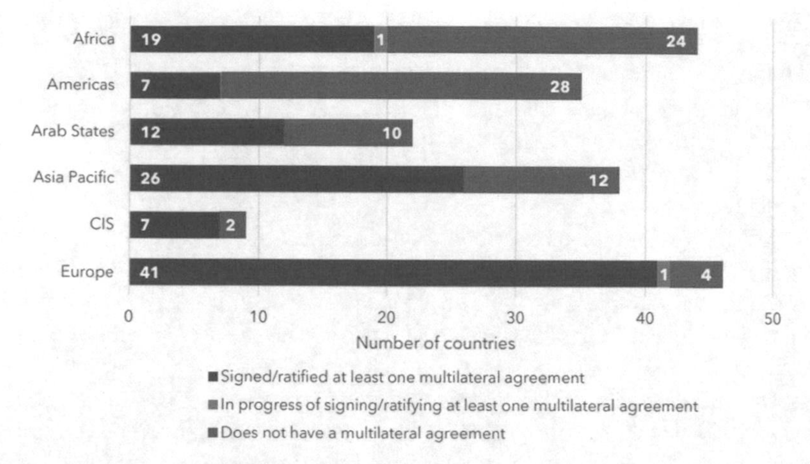

Fig. 5.4 Number of countries participating in multilateral cybersecurity agreements (signed and ratified). *Source* Global Cybersecurity Index, ITU, 2021, p. 21

The report concludes that "[c]ountries are more likely to have a multilateral agreement than a bilateral agreement, with almost 57% of countries that signed a multilateral agreement, compared to 46% of countries having signed a bilateral agreement. In addition, many countries (99) signed or ratified a multilateral agreement on information sharing and capacity development."[91]

The joint international efforts for integration of the new technologies and ensuring security are streamlined through multilateral governmental organizations, such as G7, G20, OECD, NATO, etc.

Thus, in 2018 under the presidency of Canada, the G7[92] created a Global Partnership for AI[93] (GPAI), the initiative was further facilitated during the France presidency. It grew into a multistakeholder initiative that was officially launched in 2020 and now counts 19 members. Its secretariat is accommodated by OECD, and international organization with 38 members conducting a broad number of AI-related work, e.g. consolidating information on the AI national initiatives through the AI observatory, engaging in research to advise governments, developing policy principles adopted by its Council of Ministers, which are reflected in the G-20 statement, etc. In 2019, the OECD Council adopted at the level of ministers the first intergovernmental standard on AI—the Recommendation on Artificial Intelligence (AI).[94]

[91] Global Cybersecurity Index, ITU, 2021, p. 21.

[92] Canada, France, Germany, Italy, Japan, the United Kingdom, the United States, with participation of the European Union.

[93] The global partnership on artificial intelligence, https://gpai.ai/.

[94] Recommendation of the Council on Artificial Intelligence, at https://legalinstruments.oecd.org/en/instruments/OECD-LEGAL-0449.

Global Partnership on Artificial Intelligence

"The Global Partnership on Artificial Intelligence (GPAI) is a multi-stakeholder initiative which **aims to bridge the gap between theory and practice on AI by supporting cutting-edge research and applied activities on AI-related priorities**.

GPAI's 15 founding members are Australia, Canada, France, Germany, India, Italy, Japan, Mexico, New Zealand, the Republic of Korea, Singapore, Slovenia, the United Kingdom, the United States and the European Union. They were joined by Brazil, the Netherlands, Poland and Spain in December 2020."

The collaborative work is based on the principles for *responsible stewardship of trustworthy AI* from the Recommendation of the Council on Artificial Intelligence (OECD)[95] and is implemented through the four working groups under the following themes:

1. Responsible AI
2. Data governance
3. Future of Work
4. Innovation and commercialization.

G20 "engaged in comprehensive discussion on digital economy, innovation and new industrial revolution, since 2016,[96] hold the first G20 Digital Economy Ministerial Meeting, adopted a G20 Roadmap on Digital Economy and the Ministerial Declaration. created the G20 Repository of Digital Policies and is encouraging to be used more for sharing good practices. The questions under discussion include innovative business models and faire and transparent environment, best ways to adopt new technologies benefitting economies, directions in building a new society living in the digitalized physical world (AI, 5G, IoT), ensuring free unimpeded data flows with trust. G20 based its work on a human-cantered approach and is guided by "the G20 AI Principles drawn from the OECD Recommendation on AI"—inclusiveness, human-centricity, transparency, robustness and accountability.

5.2.1 Europe Fit for Digital Age

A remarkable example was demonstrated by the **European Union** with the adoption of the first ever legal framework of AI aimed at developing an ethical, human-centric,

[95] Recommendation of the Council on Artificial Intelligence, at https://legalinstruments.oecd.org/en/instruments/OECD-LEGAL-0449.

[96] G20 Ministerial Statement on Trade and Digital Economy, p. 4. https://www.meti.go.jp/press/2019/06/20190610010/20190610010-1.pdf.

inclusive and trustworthy AI. This document is flexible and adaptive, has a risk-based approach ranging from the minimal risk, through limited, high and unacceptable risk.

In 2017, the European Commission initiated the work on developing the AI strategy. The year 2018, resulted in establishing the European High-Performance Computing Joint Undertaking[97] and the Digital Europe programme for the period 2021–2027. The same year the **European Strategy on AI** was adopted and the **Coordinated Plan on AI** published, encompassing the roadmap of action and specifying financial resources. The Member States responded with the national strategies that supported and increased investments into research and development, data generation and sharing, testing and evaluation capacities, facilitated creation and collaboration public–private partnerships, digital innovation hubs, supporting talent—doctoral and post-doctoral fellowships.[98] The General Data Protection Right (GDPR) was introduced to protect the users' privacy through warnings and choices to agree to it or not.

In 2019, European Union took further steps to ensure the new technologies correspond to the ethical standards and contribute to human rights through issuing the ethics guidelines for trustworthy AI and recommendations "Unboxing artificial intelligence: 10 steps to protect human rights".[99] It is worth noting that the High-level Expert Group on Artificial Intelligence (HLEG) working on the development of the EU ethics guidelines consisted of more than 50 representatives from academia, civil society and industry.

In 2020, HLEG developed an **Assessment List for Trustworthy AI.** The same year, the European Commission published a **White Paper** setting a vision for the AI integration within its Member States as "an ecosystem of excellence and trust" followed by a broad consultation process. The "Report on the safety and liability implications of Artificial Intelligence, the Internet of Things and robotics" provided additional explanations on technology, legislation, gaps and needs.

With the adoption of the first-ever AI legal framework[100] the European Union is aiming at becoming a "global hub for trustworthy AI". It has become an example of the legal document that embraces the new technologies providing a hospitable environment, and guarantees the people's safety, their human rights and encouraging their trust.

---

[97] Council Regulation (EU) 2018/1488 of 28 September 2018.

[98] European Union, Europe fit for the Digital Age: Commission proposes new rules and actions for excellence and trust in Artificial Intelligence, Press release, 21 April 2021, https://ec.europa.eu/commission/presscorner/detail/en/ip_21_1682.

[99] Unboxing artificial intelligence: 10 steps to protect human rights. Strasbourg: Council of Europe; 2019 (https://rm.coe.int/unboxing-artificial-intelligence-10-steps-to-protect-human-rightsreco/1680946e64, accessed 6 February 2020); Ethics guidelines for trustworthy AI. Brussels: European Commission; 2019 (https://ec.europa.eu/digital-single-market/en/news/ethics-guidelines-trustworthy-ai, accessed 13 February 2020).

[100] Regulation of the European Parliament and of the Council Laying Down Harmonised Rules on Artificial Intelligence (Artificial Intelligence Act) and amending certain Union Legislative Acts, Brussels, 21.4.2021, at https://eur-lex.europa.eu/resource.html?uri=cellar:e0649735-a372-11eb-9585-01aa75ed71a1.0001.02/DOC_1&format=PDF.

Artificial Intelligence is a fantastic opportunity for Europe. And citizens deserve technologies they can trust. Today we present new rules for trustworthy AI. They set high standards based on the different level of risks.

Today we also present a Coordinated Plan, to outline the reforms and investments we need to secure our position as a leader in #AI, worldwide. With almost €150 billion for digital investments—20% of its budget -#NextGenerationEU will help reinforce the excellence in AI.[101]

Ursula von der Leyen, President of the European Commission, Twitter, 21 April 2021

Taking into consideration the lack of desirable predictability and reliability in the innovative technologies, the legal framework classifies the AI-based technologies against four types of risks within the range from unacceptable to low. The innovations falling under the unacceptable risk are prohibited. Among them are AI systems that "manipulate human behaviour to circumvent users' free will (e.g. toys using voice assistance encouraging dangerous behaviour of minors) and systems that allow 'social scoring' by governments."[102]

High-risk applications that may endanger human life and health and bring damage require thorough evaluation before deployment, documentation, clear information to users, verification of decisions, human supervision during life-time. These are used in critical infrastructure, education, health industry (e.g., surgery), employment (selection of job applications, recruitment process), essential private and public services, law enforcement, migration, border control, etc. The systems with limited risks, such as for example chatbots, should be declared upfront to the users so that people could decide whether to continue the interaction or not. Minimal risk is presented by such systems as for example spam filters, AI-enabled video games. These systems will not be regulated, as well as those with no risk.[103]

The **Coordinated Plan on AI** has been also updated aiming to align further actions of the Member States, and increased investments (at least 1 billion EUR per year in AI-related investments in 2021–2027).

The market surveillance authorities will be supervising the implementation of the regulations. To support the oversight, develop and promote AI standards, there will

---

[101] Usula von der Leyen, Twitter, 21 April 2021, https://twitter.com/vonderleyen/status/1384815026066894849?lang=en

[102] Regulation of the European Parliament and of the Council Laying Down Harmonised Rules on Artificial Intelligence (Artificial Intelligence Act) and amending certain Union Legislative Acts, Brussels, 21.4.2021, at https://eur-lex.europa.eu/resource.html?uri=cellar:e0649735-a372-11eb-9585-01aa75ed71a1.0001.02/DOC_1&format=PDF.

[103] Regulation of the European Parliament and of the Council Laying Down Harmonised Rules on Artificial Intelligence (Artificial Intelligence Act) and amending certain Union Legislative Acts, Brussels, 21.4.2021, at https://eur-lex.europa.eu/resource.html?uri=cellar:e0649735-a372-11eb-9585-01aa75ed71a1.0001.02/DOC_1&format=PDF.

be also created a **European Artificial Intelligence Board**. In addition, regulatory sandboxes (testing environment) and voluntary codes of conduct were proposed.

The EU position to the cyber crime is reflected in the Convention **on Cybercrime** of the Council of Europe adopted in 2001 (Budapest Convention). It is "the first international treaty on crimes committed via the Internet and other computer networks"[104] and "the only binding international instrument on this issue. It serves as a guideline for any country developing comprehensive national legislation against Cybercrime and as a framework for international cooperation between State Parties to this treaty.[105]

With regards to the military application of the emerging technologies, the European Parliament resolution of 12 September 2018 on autonomous weapon systems urged "the Member States and the Council to work towards the start of international negotiations on a legally binding instrument prohibiting lethal autonomous weapon systems".[106]

The EU report on AI recognizes that "AI is causing a revolution in military doctrine and equipment through a profound change in the way armies operate, owing mainly to the integration and use of new technologies and autonomous capabilities"[107] while discussing the International Law application "suggests that research, development, and use of AI in such cases [irregular or unconventional wars] should be subject to the same conditions as use in conventional conflicts".

The report of the EU Committee on Legal Affairs on artificial intelligence
Stresses that during the use of AI a military context, Member States, parties to a conflict and individuals must at all times comply with their obligations under applicable international law and take responsibility for actions resulting from the use of such systems; underlines that under all circumstances the anticipated, accidental or undesirable actions and effects of AI-based systems must be considered the responsibility of Member States, parties to a conflict and individuals.[108]

---

[104] Details of Treaty No.185, Convention on Cybercrime, Council of Europe, https://www.coe.int/en/web/conventions/full-list/-/conventions/treaty/185?module=treaty-detail&treatynum=185.

[105] Budapest Convention and related standards, Council of Europe, https://www.coe.int/en/web/cybercrime/the-budapest-convention.

[106] 2018/2752(RSP) https://www.europarl.europa.eu/doceo/document/TA-8-2018-0341_EN.html.

[107] Report on artificial intelligence: questions of interpretation and application of international law in so far as the EU is affected in the areas of civil and military uses and of state authority outside the scope of criminal justice (2020/2013(INI)). Committee on Legal Affairs Rapporteur: Gilles Lebreton, 4.1.2021, p. 6, https://www.europarl.europa.eu/doceo/document/A-9-2021-0001_EN.pdf.

The report of the EU Committee on Legal Affairs on artificial intelligence highlighted the EU efforts aimed at LAWS being banned internationally, stressed that the LAWS debates "have failed so far", urged EU to play a leading role in regularisation of approaches to military AI, and appealed for a ban on "killer robots".[109]

5.2.2 African Digital Transformation

In 2016, the African Union established a High-Level Panel on Emerging Technologies (APET). Its members are experts in the field tasked with an advisory role to the African Union governing leadership and to its Member States "on how Africa should harness emerging technologies for economic development".[110] AI was named among the priority areas for the socio-economic development of the African continent.

> The ability to harness the power of technology and engineering to solve social problems must be accompanied by complementary adaptations in social institutions. These advances will, in turn, demand the emergence of **more scientifically and technologically enlightened societies** guided by democratic principles in the social, political, and cultural arenas."[111]
>
> *Calestous Juma, 2016*

> *Co-chair of the African Union's High-Level Panel on Science, Technology and Innovatio,*
> *Executive Secretary of the UN Convention on Biological Diversity,*
> *Founder of the African Centre for Technology Studies (ACTS)*

[108] Report on artificial intelligence: questions of interpretation and application of international law in so far as the EU is affected in the areas of civil and military uses and of state authority outside the scope of criminal justice (2020/2013(INI)). Committee on Legal Affairs Rapporteur: Gilles Lebreton, 4.1.2021, p. 7, https://www.europarl.europa.eu/doceo/document/A-9-2021-0001_EN.pdf.

[109] Report on artificial intelligence: questions of interpretation and application of international law in so far as the EU is affected in the areas of civil and military uses and of state authority outside the scope of criminal justice (2020/2013(INI)). Committee on Legal Affairs Rapporteur: Gilles Lebreton, 4.1.2021, p. 7, https://www.europarl.europa.eu/doceo/document/A-9-2021-0001_EN.pdf.

[110] The African Union High Level Panel on Emerging Technologies, AUDA-NEPAD, https://www.nepad.org/news/african-union-high-level-panel-emerging-technologies#_edn1.

[111] Juma, C. (2016). Innovation and its enemies: Why people resist new technologies. New York, NY: Oxford University Press. https://doi.org/10.1093/acprof:oso/9780190467036.001.0001.

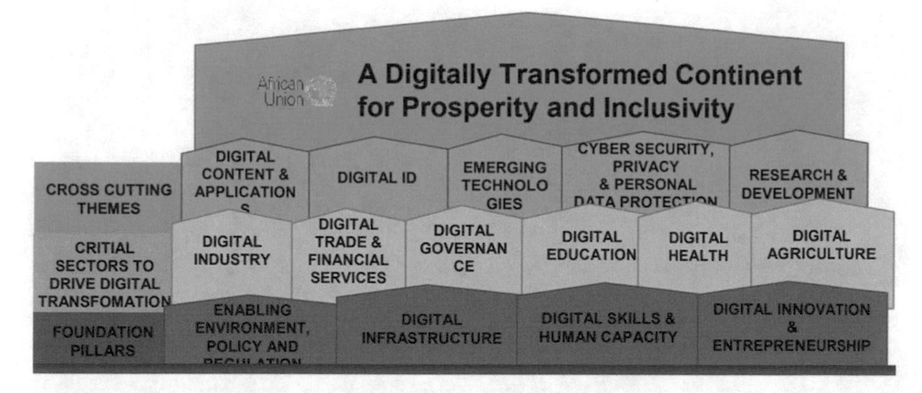

Fig. 5.5 Foundation pillars of the African digital transformation. *Source* The African digital transformation strategy (2020–2030), p. 5

In 2020, the African Digital Transformation Strategy (2020–2030)[112] (DTS) was adopted setting the vision of "Integrated and inclusive digital society and economy in Africa" and highlighting the role of innovations and digital technologies for achieving the targets of the African Union's Agenda 2063.

> Africa should not miss potentials offered by emerging technologies to transform itself and make the twenty-first century the Africa's century.[113]
>
> *The Digital Transformation Strategy for Africa (2020–2030)*

Among the strategic directions are "[b]eing prepared for digital transformation and emerging technologies such as Artificial Intelligence (AI), the Internet of Things (IoT), Machine to Machine communications (M2M) and 5G is fundamental"[114] and prepare the public policy, legal and regulatory frameworks, standards and guidance "to protect the users and society at large". The Strategy also recognizes that the emerging technologies rest unregulated in Africa and urges the governments "to keep pace" with them (Fig. 5.5).

Currently, cyber crime presents the biggest challenge for the African economy resulting in heavy losses.

---

[112] The Digital Transformation Strategy for Africa (2020–2030), https://au.int/sites/default/files/documents/38507-doc-dts-english.pdf.

[113] The Digital Transformation Strategy for Africa (2020–2030), p. 43, https://au.int/sites/default/files/documents/38507-doc-dts-english.pdf.

[114] The Digital Transformation Strategy for Africa (2020–2030), p. 7, https://au.int/sites/default/files/documents/38507-doc-dts-english.pdf.

Losses from cyber crime in Africa

Every day, large amounts of data are collected, stored and transmitted across the globe. According to Africa Cybersecurity Report 2018,[115] cybercrimes cost African economies $3.5 billion in 2017. In 2018, annual losses to cybercrimes were estimated for Nigeria at $649 million, and Kenya at $210 million. Likewise, according to the South African Banking Risk Information Centre (SABRIC), South Africa loses $157 million annually to cyber-attacks.

The continent faces a severe shortage of cybersecurity manpower. It is estimated that Africa will have a shortage of 100,000 cybersecurity personnel by 2020 As more and more economic and social activities shift into connected information spaces, volumes of trans-border data flows, specifically personal data are increasing, thus making data protection regulations critical. (p. 45).

The **African Union Convention on Cyber Security and Personal Data Protection**[116] entered into force in May 2020 establishing a legal framework in the security area. It encouraged the member States to take follow up actions and also demonstrated that they are committed to contribute to the Information Society at national, regional and international levels. It takes into account the need to protect the rights of individuals that are "guaranteed under the fundamental texts of domestic law and protected by international human rights Conventions and Treaties, particularly the African Charter on Human and Peoples' rights" (Preamble, p. 1).

Among the recent events is the establishment of the Centre of Excellence in Science, Technology and Innovation, a joint venture of the African Union Development Agency (AUDA-NEPAD), Council for Scientific and Industrial Research and Stellenbosch University. Its goal is to catalyse knowledge management, support and promote research and innovations on the continent.[117] It will also support the Science Technology Innovation Strategy for Africa developed within the African Union framework.

---

[115] SACCO [Savings and Credit Co-operative] cybersecurity report 2018, Serianu, https://www.serianu.com/downloads/SaccoCyberSecurityReport2018.pdf.

[116] The African Union convention on Cyber Security and Personal Data Protection, May 2020, https://au.int/sites/default/files/treaties/29560-treaty-0048_-_african_union_convention_on_cyber_security_and_personal_data_protection_e.pdf.

[117] Launch of AUDA-NEPAD Centre of Excellence in Science, Technology and Innovation, https://www.nepad.org/news/launch-of-auda-nepad-centre-of-excellence-science-technology-and-innovation.

5.2.3 ASEAN Digital Masterplan

In 2016, ASEAN adopted the Framework on Personal Data Protection. In 2018, the 18th ASEAN Telecommunications and Information Technology Ministers' Meeting (TELMIN) endorsed the ASEAN Framework on Digital Data Governance that set strategic goals, priorities and actions, regulate data and privacy protection, and cross-border data flows.

The ASEAN Network Security Action Council (ANSAC) has been established and discusses regional cooperation at its annual meetings. The first Digital Ministers' Meeting that took place in January 2021 adopted the ASEAN Digital Masterplan 2025.[118] It defined the vision, set the milestones in the form of outcomes and identified enabling actions.

To achieve the set goals, the countries of the region will need to ensure the infrastructural developments and digital services that will support economical and societal development, educating people to prepare experts and enhance the knowledge of non-technical experts to be able to use the new technologies.

The governance process will be equally strengthened and will include joint work of the governments, legal experts and business community complementing each other. The Masterplan requests the governing authorities to "**remov[e] unneeded regulatory barriers** to these market processes; to fund **social measures for digital inclusion and digital skills**; to **build trust** in digital services; to **harmonise regulation and standards** across ASEAN; and to **promote awareness's** of the value of digital services."[119]

It is expected that investments in the new technologies will increase, and innovations will be further stimulated. The Masterplan also encourages competition between the market players, while making specific emphasis on supporting the users' trust in the new technologies. For protection against cyber crime, the coordination and cooperation of the computer incident response teams should be strengthened.

The ASEAN Digital Masterplan 2025 envisions "ASEAN as a leading digital community and economic bloc, powered by secure and transformative digital services, technologies and ecosystem".[120]

[118] The ASEAN Digital Masterplan 2025, https://asean.org/storage/ASEAN-Digital-Masterplan-2025.pdf.

[119] The ASEAN Digital Masterplan 2025, p. 4, https://asean.org/storage/ASEAN-Digital-Masterplan-2025.pdf.

[120] The ASEAN Digital Masterplan 2025, p. 4, https://asean.org/storage/ASEAN-Digital-Masterplan-2025.pdf.

Future steps will include monitoring and evaluation measurements for secured network technologies, critical online security technologies and their deployment throughout the region, development of security indexes, framework for security in major industries, protection of critical infrastructure, collecting and sharing the best practices. Simulations exercises in cybersecurity are conducted regularly aiming at enhancing cyber preparedness and readiness, and strengthening CERT.

The above activities facilitate the development of national strategies, action plans and multisectoral coordination and governance in the AI integration.

5.2.4 United Nations Global Agenda for AI

Artificial Intelligence, with its potential to boost economic and societal development, presents an exceptional opportunity to achieve the Sustainable Development Goals (SDGs) under the United Nations Agenda for Sustainable Development. The Goal 9 of the Sustainable Development Goals aims to "[b]uild resilient infrastructure, promote inclusive and sustainable industrialization and foster innovation".

Under the auspices of the United Nations, the "AI for Good Global Summit" became a platform for global discussions on issues related to AI.

AI for Good

AI for Good is a year-round digital platform where AI innovators and problem owners learn, build and connect to identify practical AI solutions to advance the UN SDGs.

We have less than 10 years to solve the United Nations' Sustainable Development Goals (SDGs). AI holds great promise by capitalizing on the unprecedented quantities of data now being generated on sentiment behaviour, human health, commerce, communications, migration and more.

The goal of AI for Good is to identify practical applications of AI to advance the United Nations Sustainable Development Goals and scale those solutions for global impact. It's the leading action-oriented, global & inclusive United Nations platform on AI.

AI for Good is organized by ITU in partnership with 38 UN Sister Agencies, XPRIZE Foundation, ACM and co-convened with Switzerland.

https://aiforgood.itu.int/about/

The United Nations Educational, Scientific and Cultural Organization (UNESCO) is conducting broad activities related to promotion of knowledge in AI technologies, opportunities and challenges (e.g. Forum on Artificial Intelligence in Africa, December 2018, Morocco; Global conference "Principles for Artificial Intelligence: Towards a Humanistic Approach?" March 2019, France; "International Conference on Artificial Intelligence and Education" May 2019 in Beijing, China; Regional Forum on Artificial Intelligence in Latin America, December 2019, Brazil), their benefits in education, ethical standards in the AI use.

With the cyber space becoming a more critical part of our life, ensuring its peaceful use and security has become an issue of the global agenda. With full recognition of the states' sovereignty in managing the cyber space at the national level, there is no doubt that countries alone cannot cope with the increasing threats propagating globally.

According to the ITU statistics,[121] the number of the internet users increased by one billion during five years (2015–2019). The personal data, financial information have increasingly become the objects of cyber crimes.[122]

Cyber Policy Portal

The United Nations Institute for Disarmament Research (UNIDIR) recently launched a **Cyber Policy Portal** that serves as an interactive, 'at a glance' reference tool for policymakers and experts. Through a single site, users can access concise yet comprehensive cybersecurity policy profiles of all 193 UN Member States, as well as regional and international organizations and multilateral frameworks.[123]

UNIDIR organizes annually **Cyber Stability Conference**.

[121] Statistics, ITU, https://www.itu.int/en/ITU-D/Statistics/Pages/stat/default.aspx.

[122] Steve Morgan, Cybercrime to Cost the World $10.5 Trillion Annually By 2025, 13 November 2020, Special Report, at https://cybersecurityventures.com/hackerpocalypse-cybercrime-report-2016/.

[123] https://www.un.org/disarmament/update/unidir-launches-cyber-policy-portal-and-announces-date-6-june-2019-for-its-cyber-stability-conference-2019-in-new-york/.

Table 5.3 Countries with the highest cyber security index

Country name	Score	Rank
United States of America**	100	1
United Kingdom	99.54	2
Saudi Arabia	99.54	2
Estonia	99.48	3
Korea (Rep. of)	98.52	4
Singapore	98.52	4
Spain	98.52	4
Russian Federation	98.06	5
United Arab Emirates	98.06	5
Malaysia	98.06	5
Lithuania	97.93	6
Japan	97.82	7
Canada**	97.67	8
France	97.6	9
India	97.5	10

Source Global cybersecurity index, ITU, 2021[126]

The International Telecommunication Union (ITU), the United Nations specialized agency for information and communication technologies, released the fourth edition of the 2021 Global Cybersecurity Index[124] showing "a growing commitment around the world to tackle and reduce cybersecurity threats".[125] The scoring is done against five major areas, or measures as they are exercised in the country—legal, technical, organizational, capacity development, cooperation.

Table 5.3 shows the score and rank for the top ten countries with the highest cyber security index.

Cybersecurity risks are increasingly borderless, and collaboration remains an essential tool to tackle cybersecurity challenges. Cybersecurity remains a transnational issue due to the increasing interconnection and correlated infrastructures. The security of the global cyber ecosystem cannot be guaranteed or managed by any single stakeholder, and it needs national, regional, and international cooperation to extend reach and impact.

[124] Global Cybersecurity Index, ITU, 2021, at https://www.itu.int/dms_pub/itu-d/opb/str/D-STR-GCI.01-2021-PDF-E.pdf#page=33&zoom=100,120,456.

[125] https://www.itu.int/en/mediacentre/Pages/pr06-2021-global-cybersecurity-index-fourth-edition.aspx.

[126] Global Cybersecurity Index, ITU, 2021, p. 29, at https://www.itu.int/dms_pub/itu-d/opb/str/D-STR-GCI.01-2021-PDF-E.pdf#page=33&zoom=100,120,456.

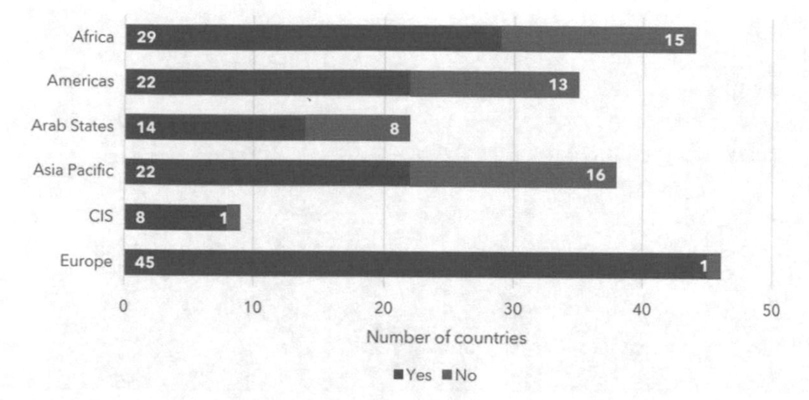

Fig. 5.6 Engagement in international activities related to cyber security. *Source* Global cybersecurity index, ITU, 2021[128]

> Bilateral and multilateral agreements are crucial in codifying norms and behaviours and enhancing international cooperation on cybersecurity.[127]
> *Global Cybersecurity Index, ITU, 2020, p. 19*

See the Fig. 5.6.

The report notes the increasing participation of countries in the international activities, highlighting that "[o]ver the past two years, 140 countries participated in international activities such as cybersecurity conferences, workshops, partnerships, and conventions with other countries."[129]

> **ITU recommendations** to promote cyber capacities
>
> To move forward, countries need to address their strengths and weaknesses in cybersecurity and leverage their competitive advantages to promote general cyber capacity and health. The Global Cybersecurity Index can help countries begin this process. To continue it, countries may need to consider:
>
> - regular assessments of their cybersecurity commitments, including meaningful metrics;
> - the continued development of national CIRTs and further establishment of sector-specific CIRTs;

[127] Global Cybersecurity Index, ITU, 2021, p. 29, at https://www.itu.int/dms_pub/itu-d/opb/str/D-STR-GCI.01-2021-PDF-E.pdf#page=33&zoom=100,120,456.

[128] Global Cybersecurity Index, ITU, 2021, at https://www.itu.int/dms_pub/itu-d/opb/str/D-STR-GCI.01-2021-PDF-E.pdf#page=33&zoom=100,120,456.

[129] Global Cybersecurity Index, ITU, 2021, p. 21, at https://www.itu.int/dms_pub/itu-d/opb/str/D-STR-GCI.01-2021-PDF-E.pdf#page=33&zoom=100,120,456.

- monitoring and updating national cybersecurity strategies with clear implementation plans;
- inclusion and diversity, especially of underrepresented groups such as women and youth, within the cybersecurity workforce;
- regular participation in international activities to share good practices, case studies, and improve preparedness and response capability;
- improving the cybersecurity capacity of micro, small, and medium-sized enterprises (MSMEs); and,
- regular engagement of all relevant stakeholders in cybersecurity, including the private sector, academia, and civil society.[130]

Global Cybersecurity Index, ITU, 2020, p. 24

The report highlighted considerable improvements in the growth of the number of computer incident response team (CIRT) (11% as compared with 2018), "[s]ome 64% of countries had adopted a national cybersecurity strategy (NCS) by year-end, while more than 70% conducted cybersecurity awareness campaigns in 2020, compared to 58% and 66%, respectively, in 2018."[131]

The report notes that many countries need to scale up training capacities for specific needs of small and medium size businesses, ensure the key area of health, energy, finance are better protected from cyber threats, secure critical infrastructure from evolving threats, guarantee data privacy. ITU encourages countries to regularly update their national cybersecurity strategies (at least every five years).

To enhance the peaceful use, security and stability in the cyber space, the work of the United Nations in this area is aimed at enabling global dialogue, developing universal rules and norms of responsible behaviour in the digital space, enhancing multilateral cooperation, including through providing platforms for negotiations, good offices and mediation services in case of conflicts, contributing to creating peaceful cyber space, ensuring confidence, building trust and accountability, respecting privacy and human rights, while being guided by the Charter of the United Nations and other international law. Multiple resolutions have been adopted to provide recommendations to Member States, e.g. the 2018 UN Resolution on "Advancing Responsible State Behavior in Cyberspace in the Context of International Security."

---

[130] Global Cybersecurity Index, ITU, 2021, p. 24, at https://www.itu.int/dms_pub/itu-d/opb/str/D-STR-GCI.01-2021-PDF-E.pdf#page=33&zoom=100,120,456.

[131] Global Cybersecurity Index, ITU, 2021, p. 4, at https://www.itu.int/dms_pub/itu-d/opb/str/D-STR-GCI.01-2021-PDF-E.pdf#page=33&zoom=100,120,456.

Report of the Group of Governmental Experts on Advancing responsible State behaviour in cyberspace in the context of international security
The Group reaffirms that an open, secure, stable, accessible and peaceful ICT environment is essential for all and requires effective cooperation among States to reduce risks to international peace and security. It is in the interest of all and vital to the common good to promote the use of ICTs for peaceful purposes. Respect for sovereignty and human rights and fundamental freedoms, as well as sustainable and digital development remain central to these efforts.[132]

Two expert working groups, the UN Groups of Governmental Experts (GGEs) on cyber security (the sixth working group since 2004, the group on Advancing responsible State behaviour in cyberspace in the context of international security) and the Open-Ended Working Group (OEWG) (since 2019) operating on consensus have been analysing the challenges (existing and potential threats) posed by the emerging technologies and digital space, developing collective solutions in consultations with Member States. Their current mandate was implemented in March[133] and May 2021[134] with the publication of two reports that reached consensus.

Report of the Group of Governmental Experts on Advancing responsible State behaviour in cyberspace in the context of international security
Incidents involving the malicious use of ICTs by States and non-State actors have increased in scope, scale, severity and sophistication. While ICT threats manifest themselves differently across regions, their effects can also be global. …. a number of States are developing ICT capabilities for military purposes; and … the use of ICTs in future conflicts between States is becoming more likely. Malicious ICT activity by persistent threat actors, including States and other actors, can pose a significant risk to international security and stability, economic and social development, as well as the safety and well-being of individuals.

In addition, States and other actors are actively using more complex and sophisticated ICT capabilities for political and other purposes. Furthermore, the Group notes a worrying increase in States' malicious use of ICT-enabled covert information campaigns to influence the processes, systems and overall

[132] Report of the Group of Governmental Experts on Advancing responsible State behaviour in cyberspace in the context of international security, Advanced copy, 28 May 2021, p. 3, at https://front.un-arm.org/wp-content/uploads/2021/06/final-report-2019-2021-gge-1-advance-copy.pdf.

[133] Open-ended working group on developments in the field of information and telecommunications in the context of international security, General Assembly, 10 march 2021, https://front.un-arm.org/wp-content/uploads/2021/03/Final-report-A-AC.290-2021-CRP.2.pdf.

[134] Report of the Group of Governmental Experts on Advancing responsible State behaviour in cyberspace in the context of international security, Advanced copy, 28 May 2021, https://front.un-arm.org/wp-content/uploads/2021/06/final-report-2019-2021-gge-1-advance-copy.pdf.

stability of another State. These uses undermine trust, are potentially escalatory and can threaten international peace and security.

New and emerging technologies are expanding development opportunities. Yet, their ever-evolving properties and characteristics also expand the attack surface, creating new vectors and vulnerabilities that can be exploited for malicious ICT activity. Ensuring that vulnerabilities in operational technology and in the interconnected computing devices, platforms, machines or objects that constitute the Internet of Things are not exploited for malicious purposes has become a serious challenge.

The Group also reaffirms that the diversity of malicious non-State actors, including criminal groups and terrorists, their differing motives, the speed at which malicious ICT actions can occur and the difficulty of attributing the source of an ICT incident all increase risk.[135]

The normative framework developed by GGE[136] is based on achievements of the previous groups and three reports published in 2010,[137] 2013[138] and 2015[139] and the work between these reports. Four pillars ground the **normative framework developed by GGE**:

1. the UN Charter and other international law are applicable in cyber space, as well as the four principles of humanity, necessity, proportionality, and distinction. This pillar is essential to maintaining peace, security and stability, open and accessible ICT environment;
2. 11 non-binding voluntary norms for the responsible state behaviour, that are built with consideration of the international law; out of them three actions to be avoided. The group concluded that "voluntary, non-binding norms of responsible State behaviour can reduce risks to international peace, security and stability".
3. Confidence-building measures aimed at strengthening transparency, predictability and stability; capacity building and cooperation measures ("to promote stability and help to reduce the risk of misunderstanding, escalation and conflict", "with a view to preserving an open, secure, stable, accessible and peaceful ICT environment.")

---

[135] Report of the Group of Governmental Experts on Advancing responsible State behaviour in cyberspace in the context of international security, Advanced copy, 28 May 2021, p. 4–5, at https://front.un-arm.org/wp-content/uploads/2021/06/final-report-2019-2021-gge-1-advance-copy.pdf.

[136] The Group started its work in 2019 following the recommendation of the General Assembly resolution 73/266.

[137] Developments in the field of information and telecommunications in the context of international security, A/65/201, 30 July 2010, https://undocs.org/A/65/201.

[138] Developments in the field of information and telecommunications in the context of international security, A/68/98, 24 June 2013, https://undocs.org/A/68/98

[139] Developments in the field of information and telecommunications in the context of international security, A/70/174, 22 July 2015, https://undocs.org/A/70/174.

4. International cooperation. Engagement of international organizations, and other stakeholders (private sector, academia, and civil society organizations)

Among the other issues discussed in the reports are protection of the critical infrastructure, safeguarding global supply chains, providing assistance when required and preventing the malicious use of digital technologies on states' national territories.

11 voluntary norms for responsible states behaviour if cyber space

Norm 13 (a) Consistent with the purposes of the United Nations, including to maintain international peace and security, States should cooperate in developing and applying measures to increase stability and security in the use of ICTs and to prevent ICT practices that are acknowledged to be harmful or that may pose threats to international peace and security.

Norm 13 (b) In case of ICT incidents, States should consider all relevant information, including the larger context of the event, the challenges of attribution in the ICT environment, and the nature and extent of the consequences.**Norm 13 (c)** States should not knowingly allow their territory to be used for internationally wrongful acts using ICTs.

Norm 13 (d) States should consider how best to cooperate to exchange information, assist each other, prosecute terrorist and criminal use of ICTs and implement other cooperative measures to address such threats. States may need to consider whether new measures need to be developed in this respect.

Norm 13 (e) States, in ensuring the secure use of ICTs, should respect Human Rights Council resolutions 20/8 and 26/13 on the promotion, protection and enjoyment of human rights on the Internet, as well as General Assembly resolutions 68/167 and 69/166 on the right to privacy in the digital age, to guarantee full respect for human rights, including the right to freedom of expression.

Norm 13 (f) A State should not conduct or knowingly support ICT activity contrary to its obligations under international law that intentionally damages critical infrastructure or otherwise impairs the use and operation of critical infrastructure to provide services to the public.

Norm 13 (g) States should take appropriate measures to protect their critical infrastructure from ICT threats, taking into account General Assembly resolution 58/199.

Norm 13 (h) States should respond to appropriate requests for assistance by another State whose critical infrastructure is subject to malicious ICT acts. States should also respond to appropriate requests to mitigate malicious ICT activity aimed at the critical infrastructure of another State emanating from their territory, taking into account due regard for sovereignty.

Norm 13 (i) States should take reasonable steps to ensure the integrity of the supply chain so that end users can have confidence in the security of ICT products. States should seek to prevent the proliferation of malicious ICT tools and techniques and the use of harmful hidden functions.

Norm 13 (j) States should encourage responsible reporting of ICT vulnerabilities and share associated information on available remedies to such vulnerabilities to limit and possibly eliminate potential threats to ICTs and ICT-dependent infrastructure.

Norm 13 (k) States should not conduct or knowingly support activity to harm the information systems of the authorized emergency response teams (sometimes known as computer emergency response teams or cybersecurity incident response teams) of another State. A State should not use authorized emergency response teams to engage in malicious international activity.[140]

Report of the Group of Governmental Experts on Advancing responsible State behaviour in cyberspace in the context of international security, 2021

This framework is considered as a beginning of further process. Further negotiations and consultations of the Member States under the auspices of the United Nations are encouraged to discuss how specific rules and principles of the international law are applicable in the context of cooperation and international security in cyber space, how to ensure higher levels of transparency and accountability through sharing views on the use of information and communication technologies. Open transparent discussion will further contribute to building confidence and the operationalization of confidence building measures.

The high importance of the cyber space security has been again reconfirmed through raising the discussion to the level of the UN Security Council and organizing the open debates. At the first ever Security Council debate on cyberthreats in June 2021,[141] the United Nations High Representative for Disarmament Affairs Izumi Nakamitsu stressed the "dramatic surge in malicious incidents in recent years, ranging from disinformation campaigns to the disruption of computer networks, contributing to diminishing trust and confidence among States" and emphasized that the covert character of cyber attacks and challenges with the attribution can escalate tensions and "encourage States to adopt offensive postures for the hostile use of these technologies". She also highlighted that the "Secretary-General's Agenda for Disarmament emphasized the need to understand and address a new generation of

[140] Report of the Group of Governmental Experts on Advancing responsible State behaviour in cyberspace in the context of international security, Advanced copy, 28 May 2021, p. 5, at https://front.un-arm.org/wp-content/uploads/2021/06/final-report-2019-2021-gge-1-advance-copy.pdf.

[141] United Nations, Security Council, Explosive' Growth of Digital Technologies Creating New Potential for Conflict, Disarmament Chief Tells Security Council in First-Ever Debate on Cyberthreats, 29 June 2021, SC/14563, https://www.un.org/press/en/2021/sc14563.doc.htm.

technology which could challenge existing legal, humanitarian and ethical norms, non-proliferation and peace and security".[142]

The debates have echoed the concerns about safety and security in cyber space, recognizing the catalysing role of the digitalized services for the economic growth. The participants highlighted the need for a broader international collaboration and cooperation, including in sharing information, collective response in tackling the use of cyber space for hostile military operations, cybercrime and cyberespionage, halting the cyber arms race, spread of fake news, and reaffirmed the commitment of the UN Member States to international law and to the framework of responsible state behaviour in cyberspace. France requested the Security Council to "oversee international peace and security in cyberspace".

5.3 Conclusion

This chapter reviewed the actions the international community is taking to integrate the untamed new technologies and emerging threats. The overview of the undertaken actions and the outcome documents of international conferences and summits shows that humanity is already well equipped with the plans of action and roadmaps that now need to be adjusted with consideration of the AI-enabled technologies being more integrated.

The adoption of the global framework on the AI-integration and governance is a necessity and it would contribute to the regulations guiding the countries. At the same time, an **international treaty on the AI research and application, monitoring and evaluation frameworks with measurements** would ensure annual countries reporting on the AI Treaty implementation, conducting regular joint simulation exercises and after-action reviews to test preparedness and readiness for any AI-related emergencies, establishing a specialized UN agency for peaceful use of AI and a global AI data centre.

Recognizing the irreversibility of the technological progress, and welcoming all innovations, humanity should be aware of accompanying challenges and establish a required level of control to ensure its own safety and security. Computerized systems and machine learning have an enormous potential, and already at this level we can see the trends of their further development.

The next chapter will present the development trends both in computer systems and AI.

[142] United Nations, Security Council, Explosive' Growth of Digital Technologies Creating New Potential for Conflict, Disarmament Chief Tells Security Council in First-Ever Debate on Cyberthreats, 29 June 2021, SC/14563, https://www.un.org/press/en/2021/sc14563.doc.htm.

References

1. Rössler B (2005) The value of privacy. Polity Press
2. Bostrom N (2014) Superintelligence: pathsm, dangers, strategies. Oxford University Press, Dangers
3. Weizenbaum J (1976) Computer power and human reason The Massachusetts Institute of Technology

Chapter 6
Prospects

Our future is a race between the growing power of technology and the wisdom with which we use it.

Stephen Hawking, Zeitgeist 2015 conference, London (March 30, 2015)

With regard to the evolutionary steps in the development of AI and machine learning, the world opinions differ—from optimistic beliefs in the human-centric ethical AI creating a technological Eden, to most pessimistic ones reflecting both the imperfection of this technology, the way we are dealing with it and emerging threats of developing a new type of alien intelligence. Between both of them there is also a middle way proposed by common sense.

One thing is absolutely clear—the AI is already on our planet accelerating all life processes exponentially. The AI research is ongoing world-wide, and collaboration is flourishing between universities, governments and private business. The number of multidisciplinary publications covering this topic has been growing.

AI leading researchers about AI prospects

At the moment, the real cutting-edge research is still being done by institutions like Google Brain, OpenAI, and DeepMind.[1]

Paul Scharre

[1] Center for a New American Security, 2017, https://www.cnas.org/.

S. Abaimov and M. Martellini, *Machine Learning for Cyber Agents*,
Advanced Sciences and Technologies for Security Applications,
https://doi.org/10.1007/978-3-030-91585-8_6

> I want us to have better AI. I don't want us to have an AI winter where people realize this stuff doesn't work and is dangerous, and they don't do anything about it.[2]
>
> *Gary Marcus*

Among the fields that profit the most from the AI implementation are human communication, health and wellness, commerce, environment protection, safety, security and the quality of our life in general. AI excels at data analysis and detection of hidden correlations, scientific and engineering research will be further boosted by AI support, promoting the creation of more advanced versions of intelligence.

This expanding application of AI will require more research, a larger data procession and higher computing capacities. Scientific thought is skyrocketing in vision, and the visionaries are already forecasting the next generation of computers and move to the quantum machine learning. This chapter will attempt to scan the horizon of future challenges for the period up to 2050 and possibly beyond.

Scanning the horizon up to 2050, we would like to highlight **two trends**. The first one is definitely related to the **technical** issues, finding solutions for the optimum way to develop technologies, ensuring their reliability, safety and security. Another one, that to our opinion is much more critical, is the need in the simultaneous **transformation of the whole society** for better integration of high technologies into our daily life. These two trends are not mutually exclusive.

Technological development

The current stage is characterized by high risks due to the imperfections of machine learning technologies and their vulnerabilities. Technical challenge consists in minimizing risks and maximizing effectiveness while striving for innovations. Scientists, designers, developers need more knowledge about what digital intelligence is, its potential and limitations, and where quantum computing may lead us.

Society transformation

New technologies with high speed and enhanced capacities for data procession, will *magnify all weaknesses and unresolved society issues* at the financial, economic, political levels. They will inevitably urge us to take concrete decisions for their governance, to update the legal frameworks for adapting to the powerful opportunities the high technologies are affording. Of specific importance will be filling in the responsibility gap, ensuring that democratic values, human rights and dignity are

[2] We can't trust AI systems built on deep learning alone, 27 September 2019, MIT Technology Review, https://www.technologyreview.com/2019/09/27/65250/we-cant-trust-ai-systems-built-on-deep-learning-alone/.

respected. If *weaponised*, they will definitely be followed by existential risks and this is up to humanity to choose which way to follow.

6.1 Technological Development

Computer systems, data storage platforms, sensors, telecommunication infrastructure, 5G, will be the major areas of progress and will be aiming to reach the same levels globally increasing connectivity and digitalization.[3]

Hybrid digitalised technologies

In the upcoming decades technology will become **hybrid** with various levels of human control. They will be more complex, powerful, interactive, co-dependent and entail the supply chain complexity. The challenges will also include their monitoring, evaluation, risk mitigation, security at local, national and international levels.

The alternative technological paths future development will require strengthening protective measures for critical infrastructure, CBRN, military operations, space communications, etc.

Risk-based standardisation and regulatory sandboxes

The development of new technologies and modernization of the old ones, will require re-evaluation of their standardisation. Innovative standardisation approaches will be used based on the risk-based approach and levels of human control over the AI-system functionality.

Verification and validation regulatory sandboxes and reference attack scenarios will become an essential part of the complex AI-system release.

A set of reference attack scenarios will be developed for all possible types of attack or specific type of system or software. During the development and deployment stages, those systems or software reference attack scenarios will be evaluated against those scenarios and the risk score will be assigned to indicate the resilience to the real-life cyber attacks, machine-learning-based or conventional. Thus, newly engineered systems will have a risk score or class.

Growing accessibility

Technologies will become cheaper, smaller, and more accessible. This will include, computers, sensors, knowledge, datasets, and open a path to innovation in miniaturised supercomputers, semiconductors and microprocessors.

---

[3] Bogan, J. and Feeney, A. (2020) Future cities: Trends and implications. Available from: https://assets.publishing.service.gov.uk/government/uploads/system/uploads/attachment_data/file/875528/Dstl_Future_Cities_Trends___Implications_OFFICIAL.pdf [Accessed 23rd September 2020].

On one hand it will attract more researchers, investors, and users and allow more widespread experimentation. On the other hand, this will fuel cybercrime, cyber attacks, information manipulation, and more impactful cyber accidents.

Combined with the growing accessibility of conventional cyber attack tools and educational resources, machine learning will become accessible as an attack tool even for people who are not technical experts.

Growing autonomy

It is expected that the sophisticated technology, following the evolution of the machine learning, will become significantly more autonomous. Devices will be able to better communicate with each other, and their swarming capacities will also increase. Machines will be able to manage themselves, without human supervision and optimise their behaviour.

Autonomous technologies will be prioritized for high-risk environment, e.g. in hazardous conditions. In space industry machine learning will be applied to satellite operations, for example, as management and coordination of the satellite constellations, with specific attention to relative positioning, communication, end-of-life management and so on.

Machine learning evolution

Machine learning algorithms will continue improving both in the direction of their explainability and reliability. The process will be enriched by the findings in neuroscience, mathematics, quantum mechanics, cognitive linguistics, psychology, etc.

It is expected that the "classical" AI will be merging with the "learning" AI, thus the hybrid systems will appear.

Trustworthy systems would become a reality through integration of neural networks (or any other machine learning model) and symbolic AI.

Neural-symbolic computing

The integration of neural models with logic-based symbolism is expected therefore to provide an AI system capable of explainability, transfer learning and a bridge between lower-level information processing (for efficient perception and pattern recognition) and higher-level abstract knowledge (for reasoning, extrapolation and planning).

Artur D'Avila Garcez, Luis C. Lamb, Neurosymbolic AI: The 3rd Wave, December 2020, p. 11, 21, https://arxiv.org/abs/2012.05876.

The research in AI algorithms will be moving towards enhancement of reasoning (spatial, physical, psychological, temporal, causal) to improve their cognitive behaviour.[4] The major research challenge will be aimed at the balancing of

[4] Gary Marcus, The Next Decade in AI: Four Steps Towards Robust Artificial Intelligence.

reasoning and learning capacities, empowering the neural models with logic and neural-symbolic computation.[5]

Balancing learning with reasoning

The aim is to identify a way of looking at and manipulating common sense knowledge that is consistent with and can support what we consider to be the two most fundamental aspects of intelligent cognitive behaviour: the ability to learn from experience and the ability to reason from what has been learned. We are therefore seeking a semantics of knowledge that can computationally support the basic phenomena of intelligent behaviour.[6]

Leslie Valiant, The 2010 Turing Award recipient

Deep learning methods will be explored further and applied wider. The question of approaches to their analysis will be among the challenges that need to be resolved.

I believe Deep Learning will be able to do anything, but we still need some conceptual breakthroughs.[7]

Geoffrey Hinton

We think that deep learning will have many more successes in the near future because it requires very little engineering by hand, so it can easily take advantage of increases in the amount of available computation and data. New learning algorithms and architectures that are currently being developed for deep neural networks will only accelerate this progress.[8]

Yann LeCunn, Yoshua Bengio, Geoffrey Hinton

Fully quantum machine learning algorithms and the adaptation of the already existing ones for the architecture of quantum computers will create quantum-specific cyber attacks. Research experimentation with quantum machine learning methods will become more widespread. Supervised and unsupervised learning will be combined in various proportions, to develop new machine learning methods and new approaches.

---

[5] I. Donadello, L. Serafini, and A. S. d'Avila Garcez. Logic tensor networks for semantic image interpretation. In IJCAI-17, pages 1596–1602, 2017.

[6] L.G. Valiant. Three problems in computer science. J. ACM, 50(1):96– 99, 2003.

[7] Interview to MIT Technology Review, EmTech conference, 2017.

[8] Yann LeCunn, Yoshua Bengio, Geoffrey Hinton, Deep Learning for AI, 2021, at https://cacm.acm.org/magazines/2021/7/253464-deep-learning-for-ai/fulltext..

In the future supervised learning will create sufficient volume of data, that those pre-trained samples could be used for the development of semi-supervised and unsupervised systems. The need for supervised machine learning will be minimal. Autoencoders, Generative Adversarial Networks, and other less known methods of combining different models will be further developed and enriched, in order to produce innovative applications of the models for data engineering purposes. New ways of sample generation will be discovered. Systems will be able to learn by themselves much better, using large volumes of already collected knowledge.

Improvement of the already existing and the development of completely new methods will lead to the evolution in the perception and capabilities of the machines. The general knowledge will increase, and the models we have about this world will be more interconnected and better adapted in dealing with unclear and uncertain data. The growing storage capacity will allow operations with larger model and significantly larger datasets. More affordable higher-density lower-energy consumption data storage, combined with constantly increasing data transfer speeds, will further push cloud systems.

Quantum and hybrid computing

It is expected that the solution will be found for quantum computer cooling and errors. Their use will be a paradigm shift in computer technologies. This will enormously increase speed in data procession, data sets creation, machine learning, cyber security.

Theoretical explorations will be aiming beyond simple adoption of quantum computers.

The quantum computing and advanced AI-systems will accelerate the development of machine learning in general, strengthening defence systems and generating more advanced malicious tools.

With the growth of cloud computing, companies will be using cloud-access to hybrid data centres, where computations will be performs by quantum or non-quantum computers, based on the complexity of the operation.

Growing cyber threats

Growing connectivity and digitalisation will increase cyber threats. New cyber attacks and physical attacks against computer systems and devices will keep emerging and bypassing older defences. Conventional attacks will keep evolving further, but the general concept of exploiting the errors is not going to be replaced any time soon.

The raise of exploitation kits as a service, ransomware as a service and botnet as a service will give further growth to Advanced Persistence Threat as a service with almost unlimited capabilities to breach any level of security.

The hybrid attacks combining cyber attacks and physical attacks to reach high-value critical targets will grow.

Machine learning will remain a niche area of expertise, with custom tools designed by skilled programmers and ICT professionals. Machine learning will be used to generate new code, including for attack purposes. The danger will be in the code, that is poorly understood by the human operators. Attacks will be stronger than expected, and damage—more severe. Generated malware can go out of control without any means to stop it.

Evolution in defence methods

Future applications of **machine learning for defence** at the local, national, and international levels will become more valuable, especially with many countries attempting digital sovereignty.

Quality of available data will improve the accuracy and flexibility of the models, while hybrid approaches will ensure the highest possible levels security. Low quality data will become widely available for training and testing of the models. However, higher quality data will become a valuable resource for the developers and researchers.

In the constant competition between signature-based and anomaly-based intrusion detection, both will win. Specifically, a **hybrid** system that uses both signature and behavioural analyses is significantly more effective in defending the perimeter of the network and the internal systems.

Combined learning methods will be used to improve the performance of machine learning in all areas of cyber security: malware analysis, network intrusion detection, and security testing.

With the hybrid tactics and technology, domains of warfare will evolve alongside.

Technological limitations of AI

The AI, like any technology, has limits, and it is reasonable to assume, that once those limits are reached, a new technology has to be used to ensure sustainable technological progress and the next industrial revolution. However, AI is far away from being fully researched and the technology is still at the early stages of development.

It is not expected that the general AI that could surpass human intelligence and go beyond will become a reality in the coming decades.

At this stage of the AI technologies, the use of machine learning for cyber security is still in its experimental stage. Human control is needed to ensure its reliability.

However, the best governance models suitable the era of high technologies may be identified with the help of the machine learning. It is always worth keeping in mind that the machines are learning form the data the humans are generating. The challenges related to the respect of ethical and moral values are equally applicable to the humans.

6.2 Societal Transformation

Society is currently challenged by the rapid spread of an overwhelmingly wide spectrum of new technologies. In each area of social life there is an innovation that has been recently made.

It is expected that more open discussions will take place globally over the changes taking place in our life, the best ways these technologies should be integrated, and the risks they are bringing. We will learn to ask difficult questions and find solutions in debates, also through global voting, identify common positions and interests, and improve our own behaviour eventually.

Among the most challenging issues will be the positioning of humans towards intelligent machines, overwhelming flows of information and new knowledge, tackling the malicious use of AI, including in society manipulation, eliminating technological gaps between rich and poor countries. Planning and setting short-term operational and long-term strategic goals in governing integration of new world-wide technologies will also be a challenge.

Human identity

Attempts in creation of artificial intelligence raise the obvious philosophical questions about the intelligence in general and what it means to be human. Engineers have the technology to create new methods, machines, and systems. But intelligence is not defined in sufficient detail to be replicated in an artificial manner.

> The actual building of a conscious machine or robot is a task for engineers, but it is essentially tied to understanding the nature of consciousness. Igor Aleksander, in 'The Engineering of Phenomenological Systems', argues that to successfully construct a machine with phenomenological awareness it is necessary to turn to philosophy to gain a better understanding of the goal of that project. However, he demonstrates that engaging in the engineering project itself affords insight on the philosophical questions about consciousness. Aleksander argues that an important aspect of building a conscious machine is building a physical machine that can move like a person. An essential aspect of being conscious of the outside world is reflexive awareness of being a thing in the world—and this is inherently related to being a physical thing situated in a particular environment. Philosophers considering the nature of consciousness have in some cases failed to fully appreciate the importance of the physical body to consciousness, so this work by engineers and scientists on the nature of consciousness delivers real philosophical insight.
>
> *Philosophy of Engineering, Volume 1, Royal Academy of Engineering*

AI, Human Brain and Ship of Theseus

"Ship of Theseus" is a classical thought experiment about the ship, replaced plank by plank due to repairs, being the same ship or not. Human body constantly replaces old cells with new ones. Suppose we invent an electronic neuron of the same size as an organic neuron, that has a function inside, that perfectly represents the behaviour of an organic human neuron. We inject this neuron into the living brain of a living person. If a single neuron is replaced with an electronic one, would that person be the same person? What if exactly 50% of the brain is replaced by the electronic neurons, evenly left and right hemispheres? What if the entire brain is replaced with the artificial neurons with the original memories?

Symbiosis between humans and AI, direct or indirect, leads to the development of new type of interaction and a new way of thinking.

Cyborg Mind

What Brain-Computer and Mind-Cyberspace Interfaces Mean for Cyberneuroethics?

With the development of new direct interfaces between the human brain and computer systems, the time has come for an in-depth ethical examination of the way these neuronal interfaces may support an interaction between the mind and cyberspace. [This bind together not just computer science and neurology, but also disciplines like] neurobiology, philosophy, anthropology and politics

To seek a path in the use of these interfaces enabling humanity to prosper while avoiding the relevant risks. the volume is the first extensive study in cyberneuroethics, a subject matter which is certain to have a significant impact in the twenty-first century and beyond.

MacKellar, Calum. Cyborg Mind: What Brain–Computer and Mind–Cyberspace Interfaces Mean for Cyberneuroethics. Berghahn Books, 2017

Man versus Machine

Once the man stops controlling the machine, the machine will start controlling the man.

Difference between the logic of man and the logic of machine can result in poor decision making from the perspective of either sides.

Marvin Minsky et al.[9] suggested to use AI only as a tool for observation and surveillance. According to them, there is a way to create a more balanced AI through the concept of "human in the loop".

One day the AIs are going to look back on us the same way we look at fossil skeletons on the plains of Africa. An upright ape living in dust with crude language and tools, all set for extinction.

Nathan Bateman, Ex Machina (2015)

Creating an artificial intelligence, equal to that of a human, but with the ability to expend, copy, and upgrade its body, never forget and easily recall any knowledge, will be the next evolutionary form of existence, transcending human limitations, but also challenging humans as species.

AI will outperform the human intellect

"Let an ultra-intelligent machine be defined as a machine that can far surpass all the intellectual activities of any man however clever. Since the design of machines is one of these intellectual activities, an ultra-intelligent machine could design even better machines; there would then unquestionably be an "intelligence explosion", and **the intelligence of man would be left far behind**. Thus, the first ultra-intelligent machine is the last invention that man need ever make, provided that **the machine is docile enough to tell us how to keep it under control**. It is curious that this point is made so seldom outside of science fiction. It is sometimes worthwhile to take science fiction seriously."

Irving John Good,[10] *1960s*

(a British mathematician, cryptologist at Bletchley Park working with Alan Turing)

[9] Minsky, M., Kurzweil, R., & Mann, S. (2013). The society of intelligent veillance. International Symposium on Technology and Society, Proceedings, 13–17. https://doi.org/10.1109/ISTAS.2013.6613095.

[10] Should we be worried about AI? Science Focus, The Home of BBC Science Focus Magazine, 17 February 2017, https://www.sciencefocus.com/future-technology/should-we-be-worried-about-ai/.

Infodemics and fake information

It is expected that the increased information flows will contain a lot of misinformation and disinformation, using methods such as DeepFakes. Its source will be virtually impossible to identify.

This will negatively affect the users and may lead to the loss of faith in the need of these inventions. If the technology is presented as mostly malicious, it will not be adopted by the mass consumers.

Furthermore, constant flow of false knowledge will negatively affect the quality of academic research, making it more complex. Speculations, guesses, myths and conspiracy theories will be replaced by intentionally false information, intelligently crafted to look like genuine results of academic research.

Privacy challenges

Privacy challenges and conspiracy theories will appear about the AI being able to collect data about every person in the world better than any human investigator.

With access to the Internet and Social Network content, AI can cross-reference live video feeds and recently published photos to find peoples' locations in real time. Furthermore, AI may use features, that are currently unknown to us, to find people even without those photos, using metadata of communications across the globe. Approach like this would solve the challenges of crime, but at the cost of any existing form of privacy, which is not always agreeable with the majority of the population.

Privacy-preserving AI has only recently started attracting attention of the researchers. However, it is not addressed nearly enough and needs a lot more attention.

The use of metadata in communication and network analysis reduces the hardware requirements, and improves both performance and privacy of the ML implementations.

Changing the mind set to probability levels in decisions

Transparency in decision-making by machines will result in assigning probability levels to each decision. Thus, people will start discussing the solutions defining probabilities. Form the "right" or "wrong", we will discuss percentage, e.g., this decision is 60% right and 40% wrong.

Similar to quantum mechanics, where the mindset of probabilities is a default way of thinking, machine learning changes the way people take decisions.

This paradigm shift is a typical example of a machine already beginning to affect a human way of making decisions.

False expectations

People expect concrete, direct and unbiased solutions and decisions from the technology that is not meant to be exact.

In a traditional logical system there is an equation, most often linear, and if the input is 2 + 2, then the outcome will always be 4. In machine learning you can trust that it will be 4 with 95% accuracy, but it also has a probability to be 5. Machine learning is closer to the laws of quantum mechanics, than to the laws of classical mechanics. that has very predictable outcomes, similar to the conventional algorithms in computers. Quantum mechanics operates with probabilities, so does machine learning.

Furthermore, machine learning is expected to learn and somewhat mimic human decision-making, and to be perfect at it. We have to keep in mind that human operators regularly make mistakes.

Bias off the past

The use of AI may cause discrimination by a great variety of criteria and features, including those that we might not even know about.

As the AI learns from the past knowledge, biases can occur due to the human history. The history of humankind is full of discrimination examples, that AI might simply use as granted.

Humans are biased by nature. Attempting to make machine free of human biases would result in a completely different behaviour.

Technological gap and technological hegemony

Technological developments need high investments. There is a probability that there will be a growing knowledge gap and technological gap between developed and developing nations in terms of access and use of AI-based technologies. Such gaps could make superpowers even more powerful, while developing countries would be left far behind. Access to predictive capabilities of machine learning will reshape the geographical map of the world.

Geopolitical tensions and threats

The merging powerful technologies are promising advantages in economic, financial, political and military areas. There is a probability, that striving for domination and defending the national interests, the AI race will be continue being escalated by great powers. In addition, the threats will increase from non-state actors.

Dependence on technologies will increase, and any offensive action will be propagating at a larger scale and with higher impact. With AI-enabled stealthy malware, the attribution challenges will be growing.

Among the contributing factors to the malicious actions will be yet weak legal environment, standardisation gaps, validation, differences in national legislation, etc.

Developing trustworthy and ethical AI jointly

It is expected that international efforts in developing safe, human centric and ethical AI will increase, supported by civil society and opinion leaders. Efforts in creating the international legal environment will become stronger, as well as requests to develop and adopt international regulations, or even a Treaty on AI.

Regional multilateral organizations will develop stronger laws and enforcement measures to become unified in the member states.

A global United Agency for AI will be created, as well as a global AI data centre.

Chapter 7
Conclusion

The book was aimed to outline the knowledge about machine learning, its conceptual and operational environment, without delving deep into mathematical analysis of the methods used. It considered the major definitions, reviewed constituting elements, machine learning process, analysed strengths and weaknesses, and future prospects. Specific emphasis in this book was made on the cyber security issues of machine learning.

The undoubtful **advantages** of the new technologies include their capacity to process big amounts of data, high speed of information processing, capacity to learn and provide solutions for complex problems while working round the clock.

However, the technical analysis and experiments show that **the technology is still immature to be fully trusted**. Machine-learning-powered cyber agents can target human operators, software and hardware with greater efficiency using conventional and emerging attacks. Machine learning has to be used for defence with great caution, as even those defences can become a vulnerable entry point for the malicious actors.

With the wide application of machine learning in all areas of our life, these powerful new technologies have vulnerabilities that can be easily exploited with malicious purposes bringing extensive damages. While recognizing the advantages of autonomous round the clock functioning security systems, we advise being cautious in applying the machine learning **intrusion detection systems** and entrusting them with cyber controls. In addition to potential errors in recognizing the adversarial attacks, they generate multiple false positive signals unnecessarily alerting human operators.

To enhance security and defence measures of AI-systems, we advise to apply a comprehensive approach at all stages of the system life cycle. Furthermore, researchers highlight the need for the development of **explainable AI**, so that people understand how the decisions are taken and eventually taking informed decisions.

Highlighting **scarcity in the academic publications** with regards to the AI offensive capabilities, the book also reviewed the machine-learning application for **offence**. It enables malware with evasive capacities, capacity to stay unrecognized

S. Abaimov and M. Martellini, *Machine Learning for Cyber Agents*,
Advanced Sciences and Technologies for Security Applications,
https://doi.org/10.1007/978-3-030-91585-8_7

until the target is reaches, making it especially dangerous for the critical infrastructure, CBRN, nuclear facilities. In addition to the imperfection of the powerful technologies, their multiple vulnerabilities and potential human errors, the AI-weaponization presents the highest risk, and the AI arms race may inevitably lead to the humanity extinction.

International community raises strong concerns against the application of the AI in weapons, especially in AWSs authorized to take life and death decisions. Among them are scientists, engineers, producers, decision-makers signing letters appealing to the highest international bodies. Internationally agreed solutions are required to guarantee peaceful use of powerful technologies, creating the legal environment contributing to their seamless integration into our life. Transparency, accountability, respect for democratic values and human rights in the technological era are issues heatedly debated in global forums.

The past decade has witnessed and explosion in the proposed and adopted AI-related regulations both at the national and international levels. The trend of creating advisory board of technical experts and oversight committees to provide advice to governments—non-technical experts taking decisions—will continue. The flexible, adaptive and scalable technologies will require a new form of governance, engaging the whole society, communities, NGOs, delegating responsibilities. The emerging new form of governance, a so-called **networked governance,** allowing the necessary level of **flexibility in learning while going and taking solutions based on peer reviews** and not on the hierarchical command matches perfectly well the needs for the integration of these technologies.

Considering the revolutionary changes that the emerging technologies are bringing into our life, it would be beneficial to initiate discussion of an **international treaty on the AI research and application**, **monitoring and evaluation frameworks with measurements**, annual countries reporting on the AI Treaty implementation, regular joint simulation exercises and after-action reviews to test preparedness and readiness for any AI-related global emergencies, establishing a specialized UN agency for peaceful use of AI and a global AI data centre.

Looking into the future and scanning the horizon up at least 2050, we brought to the reader's attention such important trend as the development of more powerful quantum computers to meet the growing requirements of machine learning in processing bigger amounts of data and resolve challenges unresolvable at current stage.

Quantum computers are a new-generation computational architecture, more powerful and completely different from the previous one. Using the simulated behaviour of even quantum particles, modern quantum computers present a fraction of what real fully quantum computers would be capable of. At the current stage, quantum computers require unique working conditions and share the main challenge with the machine learning: the potential of the new technology is yet to be fully understood and it requires extensive experimentation.

The **future challenges** have been grouped under two feasible trends: technical issues related to development of powerful safe and reliable technologies, minimizing risks and maximizing effectiveness; and—simultaneous transformation of the whole society following integration of these technologies into our life, including relations between humans and machines and self-identification of humans in the era of AI. Setting short term and long-term goals will help taking informed decisions against the milestones.

The new technologies with the high-speed potential in processing information about us, humans, are magnifying our own moral and ethical imperfections, challenging the way the society is governed, accelerating its development, changing mindset. This will definitely make us reconsider our own behaviour and way of life, inter human relations and, hopefully, find the ways to improve ourselves and create an abundant and prosperous future.

Glossary

Below is the list of selected technical terms that may be useful to understand the topic in discussion. Most of the operational definitions are provided by the authors to ease the reference for non-technical experts.

Agents

A **software agent** is a computer program implementing specific functions. Depending on the assigned functions and capacities, it may act on behalf of the user, other program or autonomously. Their types vary, as there can be single or multi-agent systems, distributed, mobile, etc.

In this book the term **cyber agent** will be used to define a software agent, that is designed for the purpose of surveillance, defence, or offence. It can be any element, device, system enabled with decision-making and/or action by cyber means (or tools), such as programmes, machine learning, etc. In its broader meaning, this definition of a cyber agent embraces "**cyber arms**—software able to function as an independent agent, **cyber tools**—cyber arms used for specific purposes, and **cyber-physical weapons**—physical weapons, i.e. robots, drones, with cyber command and control systems."[1]

Algorithms

Sets of defined step-wise programmed instructions used by computers to calculate or problem solve.

[1] Abaimov and Martellini [1], p. 1.

S. Abaimov and M. Martellini, *Machine Learning for Cyber Agents*, Advanced Sciences and Technologies for Security Applications, https://doi.org/10.1007/978-3-030-91585-8

Machine learning algorithms

Algorithms able to "learn" from a set of selected data to implement specific actions, e.g. neural networks, random trees, etc. Based on mathematical functions, they can approximate these functions using the sample data. The term "learners" is sometimes used instead.

Application

A computer program that is developed for the practical use by humans or machines, e.g. in smart phones.

Artificial Intelligence (AI)

Capacity of software to rely on and take decisions based on the provided knowledge rather than on the predefined algorithms. This term is also referred to a complex technological phenomenon that implies the ability to extract new knowledge from the available data, make decisions and enable action to reach the set goals.[2] The AI currently in use, is a so-called narrow intelligence—machine learning algorithms used for some specific tasks; the AI general intelligence potentially able to address tasks at the level of human intelligence, as well as AI superintelligence are yet the visionary expectations of a distant future. In its broader meaning, "[a]rtificial Intelligence (AI) is a fast-evolving **family of technologies** that can bring a wide array of economic and societal benefits across the entire spectrum of industries and social activities."[3]

Artificial Intelligence system (AI system)

AI system is "a system that is either software-based or embedded in hardware devices, and that displays behaviour simulating intelligence by, inter alia, collecting and processing data, analysing and interpreting its environment, and by taking action, with some degree of autonomy, to achieve specific goals."[4]

AI system lifecycle consists of "*(i)* '**design, data and models**'; which is a context-dependent sequence encompassing planning and design, data collection and processing, as well as model building; *(ii)* '**verification and validation**'; *(iii)* '**deployment**'; and *(iv)* '**operation and monitoring'**. These phases often take place in an iterative manner and are not necessarily sequential. The decision **to**

---

[2] Abaimov and Martellini [1], p. 1.

[3] Regulation of the European Parliament and of the Council Laying Down Harmonised Rules on Artificial Intelligence (Artificial Intelligence Act) and amending certain Union Legislative Acts, Brussels, 21.4.2021, p. 1, at https://eur-lex.europa.eu/resource.html?uri=cellar:e0649735-a372-11eb-9585-01aa75ed71a1.0001.02/DOC_1&format=PDF.

[4] Report on artificial intelligence: questions of interpretation and application of international law in so far as the EU is affected in the areas of civil and military uses and of state authority outside the scope of criminal justice (2020/2013(INI)). Committee on Legal Affairs Rapporteur: Gilles Lebreton, 4.1.2021, p. 6, https://www.europarl.europa.eu/doceo/document/A-9-2021-0001_EN.pdf.

retire an AI system from operation may occur at any point during the operation and monitoring phase."[5]

"*AI knowledge* refers to the skills and resources, such as data, code, algorithms, models, research, know-how, training programmes, governance, processes and best practices, required to understand and participate in the AI system lifecycle.

AI actors are those who play an active role in the AI system lifecycle, including organisations and individuals that deploy or operate AI."[6]

Autonomous Weapon System (AWS)

"A weapon system that, once activated, can select and engage targets without further intervention by a human operator."[7]

Lethal Autonomous Weapon Systems (LAWS) are "a special class of weapon systems that use sensor suites and computer algorithms to independently identify a target and employ an onboard weapon system to engage and destroy the target without manual human control of the system."[8] They are also defined as "weapons systems without meaningful human control over the critical functions of targeting and attacking individual targets."[9]

Automated systems

"A physical system that functions with no (or limited) human operator involvement, typically in structured and unchanging environments, and the system's performance is limited to the specific set of actions that it has been designed to accomplish... typically these are well-defined tasks that have predetermined responses according to simple scripted or rule-based prescriptions."[10]

Autonomy

Capacity to take-decisions and/or act independently. The level of autonomy can vary from partial to full.

---

[5] Report on artificial intelligence: questions of interpretation and application of international law in so far as the EU is affected in the areas of civil and military uses and of state authority outside the scope of criminal justice (2020/2013(INI)). Committee on Legal Affairs Rapporteur: Gilles Lebreton, 4.1.2021, p. 6, https://www.europarl.europa.eu/doceo/document/A-9-2021-0001_EN.pdf.

[6] Recommendation of the Council on Artificial Intelligence, OECD, at https://legalinstruments.oecd.org/en/instruments/OECD-LEGAL-0449.

[7] Department of Defense, Directive 3000.09, Autonomy in Weapon Systems, http://www.esd.whs.mil/Portals/54/Documents/DD/issuances/DODd/300009p.pdf.

[8] Defense Primer: U.S. Policy on Lethal Autonomous Weapon Systems, Congressional research Service, 1 December 2020, at https://fas.org/sgp/crs/natsec/IF11150.pdf.

[9] Report on artificial intelligence: questions of interpretation and application of international law in so far as the EU is affected in the areas of civil and military uses and of state authority outside the scope of criminal justice (2020/2013(INI)). Committee on Legal Affairs Rapporteur: Gilles Lebreton, 4.1.2021, p. 11, https://www.europarl.europa.eu/doceo/document/A-9-2021-0001_EN.pdf.

[10] Ilachinski [2].

Bots

The word bot is a shortened version from "robot". In cyber security, it is a malicious program that acts as a command receiver on a target system, turning one or many such systems into a botnet.

Botnet

A combination of multiple bots and a Command and Control program.

Cloud computing

Processing of information using remote (cloud) services.

Computer security

Study of the security of a specific information system.

Critical infrastructure

"Critical infrastructure constitutes basic systems crucial for safety, security, economic security, and public health of a nation. Those systems may include, but are not limited to defence systems, banking and finance, telecommunications, energy, and other".[11]

National critical information infrastructures

"Any physical or virtual information system that controls, processes, transmits, receives or stores electronic information in any form including data, voice, or video that is vital to the functioning of a critical infrastructure; so vital that the incapacity or destruction of such systems would have a debilitating impact on national security, national economic security, or national public health and safety."[12]

Cyber arms

"A cyber arm is a software able to function as an independent agent and run commands. It has a dual use nature and can be used for both offence and defence purposes."[13]

Cyber attack

A one-sided offensive activity in the computer system or network to violate confidentiality, integrity, and availability of information, disrupt services, illegally use or destroy the system.

---

[11] Global Cybersecurity Index, ITU, 2021, p. 141, at https://www.itu.int/dms_pub/itu-d/opb/str/D-STR-GCI.01-2021-PDF-E.pdf#page=33&zoom=100120456.

[12] Global Cybersecurity Index, ITU, 2021, p. 145, at https://www.itu.int/dms_pub/itu-d/opb/str/D-STR-GCI.01-2021-PDF-E.pdf#page=33&zoom=100120456.

[13] Abaimov and Martellini [1], p. 228.

Cybersecurity audits

"A security audit is a systematic evaluation of the security of an information system by measuring how well it conforms to a set of established criteria. A thorough audit typically assesses the security of the system's physical configuration and environment, software, information handling processes, and user practices. Privately managed critical infrastructures may be requested by the regulatory bodies to perform security posture assessments periodically and report on findings."[14]

Data mining
Extraction of additional knowledge from the already existing data.

Data set
A structured collection of processed data. It can be a simple text file, a database of values in a specific order, series of multiple files, or sorted records of data.

Training dataset is used for the training stage in machine learning process.
Testing dataset may be the same or different from the training dataset, and is used for the evaluation of the capacities of machine learning models.

Encoding
Transformation of data into numeric or alpha-numeric code or pattern.

Firmware
Type of software, stored in read-only memory of the electronic device, that provides low-level control over the devices it is stored on.

Intelligence
"Ability to determine the most optimum course of action to achieve its goals in a wide range of environments."[15]

Learner
See *Machine Learning algorithm.*

Machine Learning is a subfield of computer sciences, aimed at studying the structure of data and fitting that data into mathematical models that can be interpreted and further used by other programs or human operators. It is also defined as a "computer's ability to learn without being explicitly programmed".[16]

Machine learning model (*see AI system*)

A statistical model trained on the data patterns. It can be used to make predictions using new data. Models are trained using large datasets before the initial deployment,

[14] Global Cybersecurity Index, ITU, 2021, p. 147, at https://www.itu.int/dms_pub/itu-d/opb/str/D-STR-GCI.01-2021-PDF-E.pdf#page=33&zoom=100120456.

[15] 2018 UNDIR Report "The Weaponization of Increasingly Autonomous Technologies: Artificial Intelligence", as adapted from Shane Legg and Marcus Hutter, "A Collection of Definitions of Intelligence", Technical Report IDSIA-07–07, 15 June 2007, p. 9, last retrieved on 16 March 2019 at http://www.unidir.ch/files/publications/pdfs/the-weaponization-of-increasingly-autonomous-technologies-artificial-intelligence-en-700.pdf.

[16] Samuel, A. L. (1959). Some Studies in Machine Learning Using the Game of Checkers. IBM Journal of Research and Development, 3(3), 210–229. https://doi.org/10.1147/rd.33.0210.

adjusted and optimised to have the highest accuracy and precision for a particular implementation or task.

Machine learning techniques, approaches, methods

In general, these are mathematical functions used for creating and training a model.

> More specifically, they are "(a) Machine learning approaches, including supervised, unsupervised and reinforcement learning, using a wide variety of methods including deep learning;
> (b) Logic- and knowledge-based approaches, including knowledge representation, inductive (logic) programming, knowledge bases, inference and deductive engines, (symbolic) reasoning and expert systems;
> (c) Statistical approaches, Bayesian estimation, search and optimization methods."[17]

Network security

Study of communication security between systems. It focuses on the traffic and traffic flow analysis, network hardware and software behaviour, as well as on the behaviour of network communication links (ports).

Pre-processing

In data science, pre-processing is a step in the machine learning process covering the selection, filtering, refining of the collected data, and other data manipulations.

Quantum computer

A computational device based on the principles of quantum mechanics, rather than conventional transistor technology.

Qubit—quantum bit, a basic unit of quantum information.
Quantum machine learning—machine learning methods, that use quantum data.

[17] Annex 1 to Regulation of the European Parliament and of the Council Laying Down Harmonised Rules on Artificial Intelligence (Artificial Intelligence Act) and amending certain Union Legislative Acts, Brussels, 21.4.2021, Article 3.1, at https://eur-lex.europa.eu/resource.html?uri=cellar:e0649735-a372-11eb-9585-01aa75ed71a1.0001.02/DOC_1&format=PDF.

References

1. Abaimov S, Martellini M (2020) Cyber arms: security in cyberspace. CRC Press
2. Ilachinski A (2017) AI, robots, and swarms: issues, questions, and recommended studies. Center for Naval Analysis, January 2017, p 6

S. Abaimov and M. Martellini, *Machine Learning for Cyber Agents*, Advanced Sciences and Technologies for Security Applications, https://doi.org/10.1007/978-3-030-91585-8

Printed in the United States
by Baker & Taylor Publisher Services